Human Acceleration of the Nitrogen Cycle

MANAGING RISKS AND UNCERTAINTY

OECD
BETTER POLICIES FOR BETTER LIVES

This work is published under the responsibility of the Secretary-General of the OECD. The opinions expressed and arguments employed herein do not necessarily reflect the official views of OECD member countries.

This document, as well as any data and any map included herein, are without prejudice to the status of or sovereignty over any territory, to the delimitation of international frontiers and boundaries and to the name of any territory, city or area.

Please cite this publication as:
OECD (2018), *Human Acceleration of the Nitrogen Cycle: Managing Risks and Uncertainty*, OECD Publishing, Paris.
https://doi.org/10.1787/9789264307438-en

ISBN 978-92-64-30742-1 (print)
ISBN 978-92-64-30743-8 (pdf)

Co-published by IWA Publishing
Alliance House, 12 Caxton Street, London SW1H 0QS, UK
Tel: +44 (0)20 7654 5500, Fax: +44 (0)20 7654 5555
publications@iwap.co.uk
www.iwapublishing.com

ISBN: 9781789060430 (Paperback)
ISBN: 9781789060447 (eBook)

The statistical data for Israel are supplied by and under the responsibility of the relevant Israeli authorities. The use of such data by the OECD is without prejudice to the status of the Golan Heights, East Jerusalem and Israeli settlements in the West Bank under the terms of international law.

Photo credits: Illustration © Richard Wilkinson

Corrigenda to OECD publications may be found on line at: *www.oecd.org/publishing/corrigenda*.
© OECD 2018

You can copy, download or print OECD content for your own use, and you can include excerpts from OECD publications, databases and multimedia products in your own documents, presentations, blogs, websites and teaching materials, provided that suitable acknowledgement of OECD as source and copyright owner is given. All requests for public or commercial use and translation rights should be submitted to *rights@oecd.org*. Requests for permission to photocopy portions of this material for public or commercial use shall be addressed directly to the Copyright Clearance Center (CCC) at *info@copyright.com* or the Centre français d'exploitation du droit de copie (CFC) at *contact@cfcopies.com*.

Foreword

Nitrogen is one of the most important elements for life on earth. It is a key ingredient in DNA and RNA, photosynthesis and amino acids – the building blocks of proteins. Nitrogen is essential for the growth of plants and crops on which humans and livestock depend. Approximately half of the world's population rely on nitrogen fertilisers for their food consumption, making nitrogen fundamental to global food security.

Humankind has doubled the inputs of chemically-reactive forms of nitrogen to the environment since the early 20th Century. The use of nitrogen-containing fertiliser and cultivation of nitrogen-fixing crops has expanded rapidly in order to feed the planet's growing population. Nitrogen also has many industrial uses, and the combustion of fossil fuels releases nitrogen to the atmosphere.

While too little nitrogen input constrains agricultural and industrial productivity, the excessive use of nitrogen has far-reaching and complex – but often overlooked – effects. It poses a risk to human health and the environment, and threatens to undermine efforts to achieve biodiversity, climate and other Sustainable Development Goals. Once introduced into the environment, nitrogen easily changes chemical form and moves between air, soil, water and ecosystems, causing a cascade of damages. We do not yet fully understand the resilience of ecosystems to excess nitrogen, or the effects of nitrogen loading on different ecosystem services. This underscores the need for stricter monitoring and management.

This report, Human Acceleration of the Nitrogen Cycle: Managing Risks and Uncertainty – produced by the OECD in collaboration with the UNECE Task Force on Reactive Nitrogen (TFRN) – proposes a policy framework to address the environmental and public health challenges of nitrogen. It advocates a three-pronged approach to manage the known risks related to nitrogen and the uncertainties associated with excess nitrogen in the environment. This entails: (i) analysing nitrogen pathways to better manage environmental risks; (ii) addressing nitrous oxide emissions in climate change mitigation policies; and (iii) monitoring and managing residual nitrogen by measuring the effect of the above measures on the national nitrogen budget. The report provides guidance on how to implement this three-pronged approach, while addressing the need for coherence across sectoral and environmental policies.

I am proud that the OECD is working closely with the scientific community to help inform policymakers, governments, business and civil society of the importance of managing human impacts on the nitrogen cycle and promoting better nitrogen policies for better lives.

Angel Gurria

Secretary General

OECD

Acknowledgements

The overall framing and approach taken here was developed jointly by Gérard Bonnis of the Environment Directorate and Simon Buckle, Head of the Climate, Biodiversity and Water Division of the Environment Directorate, informed by many discussions with policy and academic experts working in these issues. Substantive inputs to the report were provided by: Gérard Bonnis (Chapters 1, 2, 3, 5 and Annex A); Marc Ribaudo, Economic Research Service, U.S. Department of Agriculture, Washington D.C. (Chapter 4); and Paul Drummond, Paul Ekins and Paolo Agnolucci, University College London (UCL) Institute for Sustainable Resources, London, United Kingdom (Chapter 6). Simon Upton, now the New Zealand Parliamentary Commissioner for the Environment, initiated the project during his period as Environment Director at the OECD.

The report also benefited from the presentations and discussions at the Workshop: 'The Nitrogen Cascade and Policy: Towards Integrated Solutions' jointly organised by the OECD and the UNECE Task Force on Reactive Nitrogen (TFRN), held in Paris on 9-10 May 2016 as well as a discussion at the meeting of the OECD Environment Policy Committee (EPOC) at Ministerial level (28-29 September 2016). The OECD Nitrogen Expert Group (NEG), established as an informal consultative group, provided advice on request throughout the preparation of the report.

Nassera Belkhiter assisted in the preparation of the report for publishing.

The report benefited from voluntary contributions from Korea, New Zealand and Switzerland.

Table of contents

Foreword 3
Acknowledgements 5
Abbreviations 13
Executive summary 17

Chapter 1. Why does nitrogen matter? 19
    1.1   A doubling of global nitrogen fixation since pre-industrial time 20
    1.2   An essential nutrient, but a potential pollutant 21
    1.3   Nitrogen has multiple sources 22
    1.4   ...multiple pathways 24
        1.4.1   Air 24
        1.4.2   Greenhouse balance and ozone layer 25
        1.4.3   Water 26
        1.4.4   Ecosystems and biodiversity 28
        1.4.5   Soil 29
    1.5   ...and multiple impacts 30
        1.5.1   Air quality 30
        1.5.2   Greenhouse balance and ozone layer 31
        1.5.3   Water quality 31
        1.5.4   Ecosystems and biodiversity 32
        1.5.5   Soil quality 33
    1.6   The "nitrogen cascade" 33
    Notes 36
    References 38

Chapter 2. Proposed approach to address nitrogen pollution 41
    2.1   The risk approach 43
        2.1.1   The different nitrogen risks 43
        2.1.2   Deepening pathway analysis to better manage risks of nitrogen pollution 44
        2.1.3   Feasibility of the risk approach 54
        2.1.4   The case of nitrous oxide (N_2O) 55
    2.2   The "precautionary" approach 56
    Notes 65
    References 67

Chapter 3. Examples of impact-pathway analysis and its translation into policy-making 71
    3.1   Case study 1: Impact-Pathway Analysis (IPA) and air pollution 72
        3.1.1   Urban air pollution 72
        3.1.2   Eutrophication of terrestrial ecosystems 74

3.1.3 Policy relevance of IPA for air pollution risk management 75
3.2 Case study 2: Impact-Pathway Analysis (IPA) and water pollution 80
3.2.1 Coastal water pollution 80
3.2.2 Lake water pollution 83
3.2.3 Groundwater contamination 85
3.2.4 Policy relevance of IPA for water pollution risk management 88
Notes 92
References 94

Chapter 4. The unintended consequences on the nitrogen cycle of conservation practises in agriculture 97

4.1 Managing nitrogen for agriculture and the environment 98
4.2 Nitrogen pathways in crop production 99
4.3 Nitrogen pathways in animal production 100
4.4 Conservation practices and the nitrogen cycle 101
4.4.1 Nutrient management 102
4.4.2 Tillage 103
4.4.3 Cover crops 104
4.4.4 Filter strips 104
4.4.5 Restored wetlands 104
4.4.6 Field drainage 104
4.4.7 Chemical additions to manure 105
4.4.8 Tank covers 105
4.4.9 Slurry lagoon covers 105
4.4.10 Manure incorporation and injection 105
4.5 Changing nutrient management on cropland may result in environmental trade-offs 106
4.5.1 NRCS Conservation Effects Assessment Project 107
4.6 Water-air trade-offs in manure management 108
4.7 Nitrous oxide (N$_2$O) management practices 109
4.8 Summary, conclusions and areas for further analysis 110
Notes 112
References 113

Chapter 5. Criteria to guide nitrogen policy making 119

5.1 Policy coherence 120
5.2 The effectiveness, efficiency and feasibility of policy instruments 124
5.2.1 A typology of policy instruments 124
5.2.2 Effectiveness, efficiency and feasibility criteria 126
5.3 Unintended effects related to the nitrogen cascade 129
Notes 132
References 133

Chapter 6. An assessment of the effectiveness, efficiency and feasibility of nitrogen policy instruments 137

6.1 Key findings 138
6.2 Case studies of policy instruments 146
6.2.1 The Swedish refund emission payment for nitrogen oxides (NO$_x$): a combination of environmentally related tax and public financial support (PFS) 146
6.2.2 Selected instrument combinations of relevance to nitrogen pollution 147

6.2.3	The Greater Miami Watershed Trading Programme: an example of tradable permit system (TPS)	151
6.2.4	Japan's automobile 'nitrogen oxides (NO_x) law': an example of direct environmental regulation (DER)	152
6.2.5	Pennsylvania's Resource Enhancement and Protection Programme: an example of public financial support (PFS)	154
6.2.6	The Agriculture and Environment Programme for Vittel area: an example of payment for ecosystem services (PES)	154
6.2.7	Australia's 'FERTCARE': an example of information measure	155
6.2.8	Chesapeake 2000 Programme: an example of voluntary scheme	157

Notes 159
References 161

Annex A. Basic facts on nitrogen 165

A.1	The nitrogen cycle	165
A.2	The nitrogen problem in brief	166
A.3	Supplementary information on nitrogen impacts	167
A.3.1	Air quality	167
A.3.2	Greenhouse balance	169
A.3.3	Water quality	169
A.3.4	Ecosystems and biodiversity	170

Notes 172
References 173

Tables

Table 1.1. Contributions to annual global nitrogen fixation 20
Table 1.2. Key threats of excessive release of nitrogen into the environment 22
Table 1.3. Anthropogenic sources of nitrogen 22
Table 2.1. A three-pronged approach to address nitrogen pollution 43
Table 2.2. Planetary boundaries for anthropogenic nitrogen fixation 62
Table 3.1. Estimated direct deposition of nitrogen to the tidal Chesapeake Bay 83
Table 3.2. Estimated nitrogen loads in the Chesapeake Bay 83
Table 3.3. Sources of river nitrogen exports to coastal waters 91
Table 4.1. Percentage of cropland meeting nitrogen rate, method, and timing criteria consistent with "good" management, Mississippi River Basin, 2003-06 98
Table 4.2. Changes in nitrogen loss to the atmosphere with the adoption of nutrient management practices, by major watershed 107
Table 4.3. Practices to reduce nitrous oxide (N_2O) emissions 110
Table 5.1. Fertiliser subsidies in selected BRIICS countries 121
Table 5.2. Greenhouse gas (GHG) emissions related to use and production of nitrogen fertilisers 121
Table 6.1. Effectiveness, efficiency and feasibility of policy instruments 140
Table 6.2. Effectiveness, efficiency and feasibility of policy instrument combinations 143

Figures

Figure 1.1. Feeding the world with artificial fertiliser 21
Figure 1.2. Nitrogen emission and deposition pathways 25

Figure 1.3. Sources of stratospheric nitrogen oxides (NO_x) 26
Figure 1.4. Nutrient pathways in freshwater 27
Figure 1.5. Nutrient pathways in marine water 28
Figure 1.6. Nitrogen pathways in soil 30
Figure 1.7. The nitrogen cascade 34
Figure 1.8. Example of serial impacts on health and the environment of a single nitrogen atom 35
Figure 2.1. Nitrogen emission zones for different pollution risks 46
Figure 2.2. Sources and pathways of nitrogen in South and Central Puget Sound 47
Figure 2.3. State of nitrogen regulation in emission zones 49
Figure 2.4. Nitrogen flows and cascading damage costs in the Chesapeake Bay watershed 51
Figure 2.5. Costs of damage and reduction of different forms and sources of nitrogen in the San Joaquin Valley, California 52
Figure 2.6. Global average atmospheric concentrations of nitrous oxide (N_2O) 56
Figure 2.7. Aquatic systems at risk of nitrogen pollution in central California 59
Figure 3.1. Particle alert threshold exceeded in Paris in March 2014 73
Figure 3.2. Nitrogen deposition loads in Baden-Württemberg, 2009 74
Figure 3.3. Nitrogen dioxide (NO_2) exceeds legal limits in many EU cities despite the reduction of nitrogen oxides (NO_x) emissions at national level 76
Figure 3.4. Particulate matter (PM) concentrations exceed and are projected to continue to exceed legal limits in a business as usual scenario 77
Figure 3.5. Urban population exposure to ground-level ozone (GLO) is and is projected to remain a concern in a business as usual scenario 79
Figure 3.6. Chesapeake Bay's criteria for dissolved oxygen 80
Figure 3.7. Chesapeake Bay airsheds for nitrogen oxides (NO_x) and ammonia (NH_3) 81
Figure 3.8. Chesapeake Bay Programme modelling framework 82
Figure 3.9. Nitrogen transport times in the Rotorua Lake basin 84
Figure 3.10. Hydrology of the Willamette Basin 87
Figure 3.11. Global distribution of ocean dead zones 89
Figure 3.12. OECD agricultural nitrogen: fewer surpluses and more efficiency in use 90
Figure 5.1. Policy assessment criteria 120
Figure 5.2. Greenhouse gas (GHG) emissions of crops by soil type 122
Figure 5.3. China and OECD: trends in level and structure of agricultural support 123
Figure 5.4. New Zealand's greenhouse gas (GHG) Emissions Trading Scheme 124
Figure 6.1. Design of a 'feebate' instrument against a tax/charge 149

Figure A.1. The nitrogen cycle: major processes 165

Boxes

Box 2.1. The social costs of a kilogram of nitrogen fertiliser 53
Box 2.2. Differentiating between risk and uncertainty 57
Box 2.3. Estimating boundaries for different nitrogen forms 61
Box 2.4. The International Nitrogen Management System 63
Box 3.1. Groundwater recharge pathways in the Willamette Basin, Oregon 87
Box 3.2. A compelling evidence of the rapid increase in ocean dead zones 88
Box 4.1. Conservation practices are often used in combination 102
Box 5.1. Example of positive side effect of environmental policies primarily aimed at reducing nitrogen pollution 128

Box 5.2. The synergy effect on nitrous oxide (N_2O) emissions of practices to remove nitrate (NO_3^-) from sewage 130

Abbreviations

AGREV	Agriculture and Environment Programme for Vittel area
Anammox	Anaerobic ammonium oxidation
BMP	Best management practices
BRIICS	Brazil, Russia, India, Indonesia, China and South Africa
CEAP	Conservation Effects Assessment Project
CMAQ	Community Multi-scale Air Quality (model)
CO$_2$	Carbon dioxide
DER	Direct environmental regulation
DNRA	Dissimilatory nitrate reduction to ammonium
DON	Dissolved organic nitrogen
ERS	Economic Research Service (United States Department of Agriculture)
EU	European Union
EUR	Euro
GDP	Gross Domestic Product
GEF	Global Environment Fund
GHG	Greenhouse gas
GLO	Ground-level ozone
GMWTP	Greater Miami Watershed Trading Programme
GPNM	Global Partnership on Nutrient Management (UNEP)
GWAVA	Groundwater vulnerability and assessment (model)
GWh	Gigawatt hour
GWP	Global Warming Potential
HNO$_3$	Nitric acid
IFA	International Fertiliser Association
InMAP	Intervention Model for Air Pollution
INMS	International Nitrogen Management System

IPA	Impact-Pathway Analysis
IPCC	Intergovernmental Panel on Climate Change
KWh	Kilowatt hour
LRTAP	Convention on Long-range Transboundary Air Pollution
N_2	Dinitrogen (the inert form of nitrogen)
N_2O	Nitrous oxide
N_2O_5	Dinitrogen pentoxide
NANI	Net anthropogenic nitrogen inputs
NH_3	Ammonia
NH_4^+	Ammonium
NH_4NO_3	Ammonium nitrate
NH_x	Reduced forms of nitrogen (includes ammonia and ammonium)
Nitrogen	Reactive forms of nitrogen (excludes N_2)
Nitrogen cycle	Biogeochemical cycle of nitrogen
NLEAP	Nitrogen Loss and Environmental Assessment Package (model)
NO	Nitric oxide
NO_2	Nitrogen dioxide
NO_2^-	Nitrite
NO_3^-	Nitrate
NO_x	Nitrogen oxides (includes nitric oxide and nitrogen dioxide)
NO_y	Oxidised forms of nitrogen (includes nitrogen oxides, nitric acid and dinitrogen pentoxide)
NRCS	Natural Resource Conservation Service (United States Department of Agriculture)
NUE	Nitrogen Use Efficiency
O_3	Ozone
P	Phosphorus
PAN	Peroxyacetylnitrate
PAS	Integrated Approach to Nitrogen (Netherlands)
Pathways	Biogeochemical pathways
PB	Planetary boundaries

PES	Payment for ecosystem services
PFS	Public financial support
PON	Particulate organic nitrogen
PM	Particulate matter
PM$_{2.5}$	Fine particles (1-2.5 micrometres in diameter)
PM$_{10}$	Coarse particles (1-10 micrometres in diameter)
PP	Precautionary principle
PRMS	Precipitation-Runoff Modelling System (model)
REAP	Resource Enhancement and Protection Programme
SEK	Swedish krona
SWV-GWMA	Southern Willamette Valley groundwater management area (Oregon, United States)
TFRN	Task Force on Reactive Nitrogen (UN Economic Commission for Europe)
TMDL	Total Maximum Daily Load
TPS	Tradable permit system
UN	United Nations
UNEP	United Nations Environment Programme
USD	United States dollar
VNT	Vintage Nitrogen Trading (New Zealand)
VS	Voluntary scheme
WFD	Water Framework Directive (EU)
WHO	World Health Organisation (UN)
WQSTM	Water Quality and Sediment Transport Model (model)
WTO	World Trade Organisation
WWTP	Wastewater treatment plant

Executive summary

Nitrogen is often the limiting nutrient for the growth of plants and crops on which animals and humans feed. The rapid growth in the use of fertilisers has been one factor contributing to increased crop yields. Around half the world's population depends on nitrogen fertilisers for their food consumption, making nitrogen essential to global food security, and it will be increasingly so as population grows to an estimated 9.7 billion by 2050.

Humans have doubled the annual flow of nitrogen following the discovery of the Haber-Bosch process a century ago, which allowed us to convert molecular nitrogen from the air into usable nitrogen for fertiliser (80%) and industrial uses (20%). The expansion of cropland devoted to nitrogen-fixing legumes and the large-scale burning of fossil fuels have also increased nitrogen fixation. The relative size of the modification of the nitrogen cycle by human activity is far greater in magnitude than the parallel modification of the carbon cycle.

Nitrogen moves among environmental media and takes on multiple forms, creating multiple risks to the environment. Atmospheric emissions of nitrogen oxides reduce air quality via the creation of ground-level ozone and, when combined with ammonia, via particulate matter, increasing human health risks such as respiratory illnesses and cancer. Nitrates in water bodies contribute to eutrophication in lakes and coastal areas, impacting on fisheries, and affecting drinking water quality. Nitrogen can also damage ecosystems through acidification of soils and seas.

Nitrogen is also intimately linked to climate change through its influence on rates of biological activity and the uptake of carbon dioxide by ecosystems, influencing the so-called carbon fertilisation effect. Additionally, nitrous oxide is an important greenhouse gas in its own right. In the stratosphere, nitrous oxide is also a powerful ozone depleting substance.

Each of these nitrogen forms contributes to a sequence of environmental and human health impacts. This has been referred to as the nitrogen 'cascade'. Uncertainties abound concerning such cascading effects. We do not fully understand the resilience of ecosystems to increased nitrogen loading. We are often unaware of the ecosystem services we may be losing by an insufficient focus on the wider consequences of the human acceleration of the nitrogen cycle.

Such uncertainties, coupled with the need to manage the risks of nitrogen pollution and to monitor and control the steady increase in atmospheric nitrous oxide concentrations, calls for a three-pronged approach:

- First, there is a need to manage local pollution risks by better understanding the pathways of nitrogen between sources and impacts (the "spatially targeted risk approach")

- Second, account must be taken of the observed increase in global atmospheric concentrations of nitrous oxide that have consequences for both climate change and stratospheric ozone (the "global risk approach")
- Third, there is a need to prevent "excessive" nitrogen entering the environment by developing strategies addressed to the different sources (on the basis of the most cost effective means) to reduce them (the "precautionary approach").

Regardless of the approach (risk or precautionary), evaluation criteria are needed to select the right nitrogen policy instruments. First, it is necessary to assess and manage the unintended effects on nitrogen emissions of policies primarily aimed at economic objectives (agricultural production, energy supply) or environmental objectives other than nitrogen pollution (e.g. climate change). Secondly, nitrogen policy instruments can then be selected on the basis of their cost-effectiveness and provided that the "feasibility" of their implementation is not a problem.

To be effective, nitrogen policy measures should also consider possible unintended effects due to the nitrogen cascade. In particular, efforts to lessen the impacts caused by nitrogen in one area of the environment should not result in unintended nitrogen impacts in other areas ("pollution swapping" effects), and should maximise opportunities to reduce other nitrogen impacts ("synergy" effects).

Chapter 1. Why does nitrogen matter?

This chapter explains why the nitrogen cycle is an important issue for environmental policy. It provides an overview of the main sources of nitrogen, the pathways normally used by nitrogen once released into the environment, and the health and environmental risks associated with excess nitrogen in the receiving ecosystems. The chapter introduces the concept of "nitrogen cascade", which translates an unpredictable sequence of nitrogen cycle effects.

Three elements, carbon, oxygen and hydrogen, constitute more than 90% of our bodies by weight. Carbon comes from carbon dioxide (CO_2), which plants take in by photosynthesis and turn into food. We breathe atmospheric oxygen into our blood, and hydrogen reaches us through water. But the most important element in many ways for humans is the 4th most common in our bodies – and the hardest to find in nature in forms we can use: nitrogen.[1] Every living creature requires nitrogen. For example, nitrogen is a major component of chlorophyll, the most important pigment needed for photosynthesis. It is required in all the DNA and RNA in our cells. It is also required in amino acids, the key building blocks of proteins.

The atmosphere we breathe is 78% dinitrogen.[2] However, if we had to rely on dinitrogen only, it would be like floating on the sea, dying of thirst (Hager, 2009). This is because dinitrogen in the atmosphere is in the form of N_2 molecules which have a very strong (triple) chemical bond and are basically useless to living creatures in that form. Conversion of dinitrogen to biologically available ("reactive") nitrogen, a process called "fixation", involves several processes, as part of the nitrogen cycle (Annex A).

1.1 A doubling of global nitrogen fixation since pre-industrial time

Until the end of the 19th century, limited availability of nitrogen severely constrained agricultural and industrial productivity.[3] At the start of the twentieth century, several industrial processes were developed to fix dinitrogen into nitrogen. The most important process turned out to be the Haber–Bosch process.[4] Prior to the Haber–Bosch discovery, in 1913, nitrogen fixation depended almost entirely on bacteria (see Annex A). By adding industrial clout to the efforts of the microbes that used to do the job single-handed, humans have dramatically increased the annual amount of nitrogen fixed on land (The Economist, 2011). Humans have actually doubled the annual input of nitrogen (Table 1.1).[5] By comparison, our intervention in the carbon cycle has added roughly 10% to the natural pre-industrial land-atmosphere flux (Prentice et al., 2015).

Table 1.1. Contributions to annual global nitrogen fixation

Mechanism	Amount[1] (million tonnes N per year)
Terrestrial pre-industrial biological fixation	58
Marine biological fixation	140
Lightning fixation of nitrogen	5
Sub-total of "natural" fixation	203
Biological fixation by croplands	60
Combustion	30
Fertiliser and industrial feedstock	120
o/w fertiliser	96
o/w industrial feedstock[2]	24
Sub-total of "anthropogenic" fixation	210
Total fixation	413

1. Many uncertainties remain about the stocks and flows of nitrogen within and between air, land, freshwater and oceans. For example, a "remarkably large uncertainty" remains as to the extent of biological fixation of nitrogen (Stocker et al., 2016) and recent studies show that more than a quarter of the nitrogen used by plants would come from the Earth's bedrock (Houlton et al., 2018).
2. Including nylon, plastics, resins, glues, melamine, animal/fish/shrimp feed supplements, and explosives.
Source: Fowler et al. (2013).

1.2 An essential nutrient, but a potential pollutant

The Haber–Bosch process is the main industrial process for the production of nitrogen. Most (80%) is used to produce fertiliser; the remaining 20% is used to make explosives, dyes, household cleaners and nylon. Erisman et al., 2008 estimated that around half of the nitrogen in the protein that humans eat today got into that food by way of artificial fertiliser (Figure 1.1), illustrating the world's increasing dependence on the use of fertiliser.

Figure 1.1. Feeding the world with artificial fertiliser

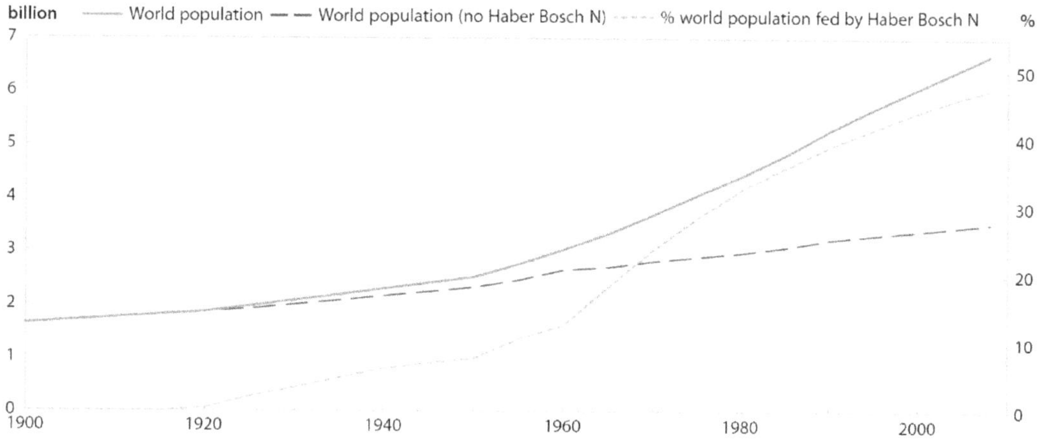

Note: The Haber–Bosch process was discovered in 1913.
Source: Erisman et al. (2008).

Today, nearly three-quarters of anthropogenic nitrogen is intentionally added to soils by cultivation of nitrogen-fixing crops and application of chemical fertiliser. An additional 10% is fixed intentionally to produce industrial feedstock and the remaining 15% is unintentionally created as nitrogen oxides (NO_x) into the air as a result of energy combustion.[6] Despite immense benefits in terms of food and energy security, losses of nitrogen fertiliser and nitrogen from combustion in the environment have numerous and significant side effects on human health and the environment, as evidenced by various recent assessments (Sutton et al., 2011; USEPA-SAB, 2011; SRU, 2015; Tomich et al., 2016; Abrol et al., 2017; Hellsten et al., 2017). The top five threats of excessive release of nitrogen into the environment are water quality; air quality; greenhouse balance and ozone layer; ecosystems and biodiversity; and, soil quality, which Sutton et al., 2011 summarises by the acronym WAGES (Table 1.2).

Table 1.2. Key threats of excessive release of nitrogen into the environment

Environmental issue	Adverse impact on health and the environment	Main form of nitrogen involved
Water	Nitrate contamination of ground- and drinking water	Nitrate (NO_3^-)
Air	Human health effects and effects on vegetation	Nitrogen oxides (NO_x), ammonia (NH_3), particulate matter (PM)[1], ground-level ozone (GLO)
	Effects on materials and monuments	Nitric acid (HNO_3), PM[1], GLO
Greenhouse balance and ozone layer	Global warming and ozone layer depletion	Nitrous oxide (N_2O)
Ecosystems and biodiversity	Eutrophication and acidification of terrestrial ecosystems	NO_3^-, ammonium (NH_4^+)
	Eutrophication of freshwater and marine ecosystems	NO_3^-, organic nitrogen[2]
Soil	Acidification	Organic nitrogen[2]

1. Coarse particles (1-10 micrometres in diameter, or PM_{10}) and fine particles (1-2.5 micrometres in diameter, or $PM_{2.5}$).
2. Dissolved and particulate organic nitrogen.
Source: Adapted from Geupel (2015) and Sutton et al. (2011).

1.3 Nitrogen has multiple sources

Table 1.2 summarises the main reactive forms of nitrogen. Human activities often represent a significant part of their emissions (Table 1.3).

Table 1.3. Anthropogenic sources of nitrogen

Environmental impact	Share of anthropogenic emissions in total emissions	Activity
Water quality	NO_3^- (?)	Agriculture, urban and industrial sewage, atmospheric deposition
Air quality	NO_x (~ 70%)	Burning fossil fuels and, to a lesser degree, biomass (the latter mainly due to slash-and-burn agriculture)
	NH_3 (~ 90-100%)	Agriculture (volatilisation following spreading of livestock manure or urea fertiliser)
	PM (?)	Formed in the atmosphere from precursors NO_x and NH_3
	GLO (~ 80%)	Formed in photochemical processes from precursor NO_x
Greenhouse balance and ozone layer	N_2O (~ 40%)	Agriculture and, to a lesser degree, burning fossil fuels and biomass, industrial processes, atmospheric deposition and sewage
Ecosystems and biodiversity	NH_3 and NH_4^+ (~ 90-100%)	Agriculture (volatilisation following spreading of livestock manure or urea fertiliser)
	NO_3^- and organic nitrogen (?)	Agriculture, urban and industrial sewage, atmospheric deposition
Soil quality	NH_4^+ (~ 90-100%); NO_3^- and organic nitrogen (?)	Agriculture and, to a lesser degree, atmospheric deposition

Source: OECD Secretariat.

Human activity is a major source of NO_x in the troposphere, especially the burning of fossil fuels.[7] In such combustion processes, nitrogen from the fuel or dinitrogen from the air combines with oxygen atoms to create nitric oxide (NO). This further combines with ozone (O_3) to create nitrogen dioxide (NO_2). The initial reaction between NO and O_3 to form NO_2 occurs fairly quickly during the daytime, with reaction times on the order of minutes. Due to the close relationship between NO and NO$_2$, and their ready interconversion, these species are often grouped together and referred to as NO$_x$. The majority of NO$_x$ emissions are in the form of NO. For example, 90% or more of tail-pipe NO$_x$ emissions are emitted as NO (Richmond-Bryant et al., 2016). Actually, only a small part of vehicle NO$_x$ emissions come from the oxidation of organic nitrogen in gasoline or diesel fuel.[8] Vehicle NO$_x$ comes mainly from incoming air that mixes with fuel inside the vehicle's engine.[9]

Human activity is also a major source of NH$_3$ emissions. Global NH$_3$ emissions into the atmosphere are dominated by agricultural practices (e.g. in Western Europe and the United States, as much as 90–100% of NH$_3$ emissions result from agriculture).[10] Agricultural NH$_3$ is emitted mainly by manure (excretion, storage and spreading) and, to a lesser extent, through volatilisation following the application of mineral fertilisers, especially when temperatures are mild and soils are moist and warm.

A range of nitrogen forms are created in the atmosphere from the oxidation of NO and NO$_2$.[11] The main pathway is oxidation to nitric acid (HNO$_3$).[12] HNO$_3$ can react with NH$_3$ to form ammonium nitrate (NH$_4$NO$_3$), a secondary PM known as "fine nitrate aerosol",[13] or be absorbed onto primary PM$_{10}$ (e.g. dust, sea salt) to form "coarse nitrate aerosols". NH$_4$NO$_3$ is also formed by reaction in liquid phase (e.g. droplets) of HNO$_3$ with NH$_4^+$ and subsequent water evaporation. In the atmosphere, NH$_3$ reacts not only with HNO$_3$ but also with other acidic gases such as sulphuric acid to form ammonium sulphate aerosol and hydrochloric acid to form ammonium chloride aerosol.

GLO is not directly emitted. It is a secondary (novel) pollutant formed by sunlight driven atmospheric chemical reactions involving carbon monoxide (CO), volatile organic compounds (VOCs), including methane (CH$_4$), and NO$_x$. Contrary to NO$_x$, there is substantial background GLO pollution. In the United States, background GLO — defined as stratosphere-to-troposphere transport,[14] imports of GLO and GLO from natural sources[15] — contributes a major portion to estimated total GLO (Lefohn et al., 2014).

Natural sources create some 60% of total N$_2$O emissions, in particular from microbial denitrification in soils under natural vegetation and in the oceans (Denman et al., 2007). Anthropogenic sources account for only 40% of total emissions, with agriculture accounting for about two thirds of anthropogenic emissions (ibid). Agricultural N$_2$O is emitted directly from fertilised agricultural soils and livestock manure (some 60%) and indirectly from runoff and leaching of nitrogen fertilisers (the remaining 40%). Other anthropogenic sources are fossil fuel combustion and industrial processes, biomass burning, atmospheric deposition and, to a lesser extent, human sewage. N$_2$O is a by-product of fossil fuel combustion in mobile and stationary sources. Industrial processes also emit N$_2$O, particularly during the production of nitric acid (an important ingredient for synthetic fertilisers) and adipic acid (primarily used for making synthetic fibres). Biomass burning (e.g. to destroy crop residues or clear land for agricultural or other uses) oxidises part of the nitrogen that is in the biomass and surrounding air, creating N$_2$O emissions. Atmospheric deposition provides terrestrial and aquatic ecosystems with extra nitrogen, which stimulates microbial denitrification. N$_2$O is also released from the bacteriological treatment (denitrification) of waste water.

In addition to agriculture and wastewater (urban and industrial), another important source of nitrogen in water is atmospheric deposition. Seitzinger et al., 2010 divides the main sources of nitrogen exports to the oceans between agriculture (50%), natural biological fixation (25%), atmospheric deposition (20%) and wastewater (5%). These data are global averages, the percentages varying according to coastal areas.

Common inorganic nitrogen forms in water are NO_3^-, nitrite (NO_2^-), NH_4^+ and NH_3. NO_3^- is stable: it is highly soluble (dissolves easily) and is easily transported in streams and groundwater. NO_2^- is relatively short-lived and is quickly converted to NO_3^- by bacteria. NH_3 is the least stable form of nitrogen in water: it is easily transformed to NO_3^- in waters that contain oxygen and can be transformed to nitrogen gas in waters that are low in oxygen. The dominant form between NH_4^+ and NH_3 depends on the pH and temperature of the water. A significant part of the total nitrogen load in surface waters is in the form of organic nitrogen, including dissolved organic nitrogen (DON) and particulate organic nitrogen (PON), even in NO_3^- enriched rivers. There is uncertainty as to how quickly organic nitrogen (DON and PON) is being recycled through the food web. For example, DON tends to be stored for much longer periods of time (up to 200 years) in high-elevation lake ecosystems (Goldberg et al., 2015).[16]

1.4 ...multiple pathways

1.4.1 Air

Apart from N_2O which is a very stable molecule in the troposphere[17] where it has a greenhouse effect, other atmospheric emissions of nitrogen can undergo rapid changes and create health hazards in the troposphere. They then deposit on the earth's surface within hours or days, creating a risk to ecosystems[18] (Salomon et al., 2016). They impact in part on the area near the source of emissions, but some nitrogen emissions may be transported over considerable distances before being deposited and having harmful effects. This is because both NH_3 and NO_x can react with each other or with atmospheric components, forming or attaching to aerosol particles which can be transported for thousands of kilometres before they are removed from the atmosphere in the process known as wet deposition, i.e. mainly by precipitation (Hertel et al., 2011).

NO_x and NH_3 can undergo chemical and physical transformation as they disperse from their source, leading to different forms of deposition (Figure 1.2). For oxidised nitrogen,[19] these comprise dry deposition of NO_x gases and wet deposition of NO_3^- while for reduced nitrogen, these comprise dry deposition of NH_3 gas and wet deposition of NH_4^+. Dry deposition of particulate and aerosol nitrate and ammonium can also contribute.

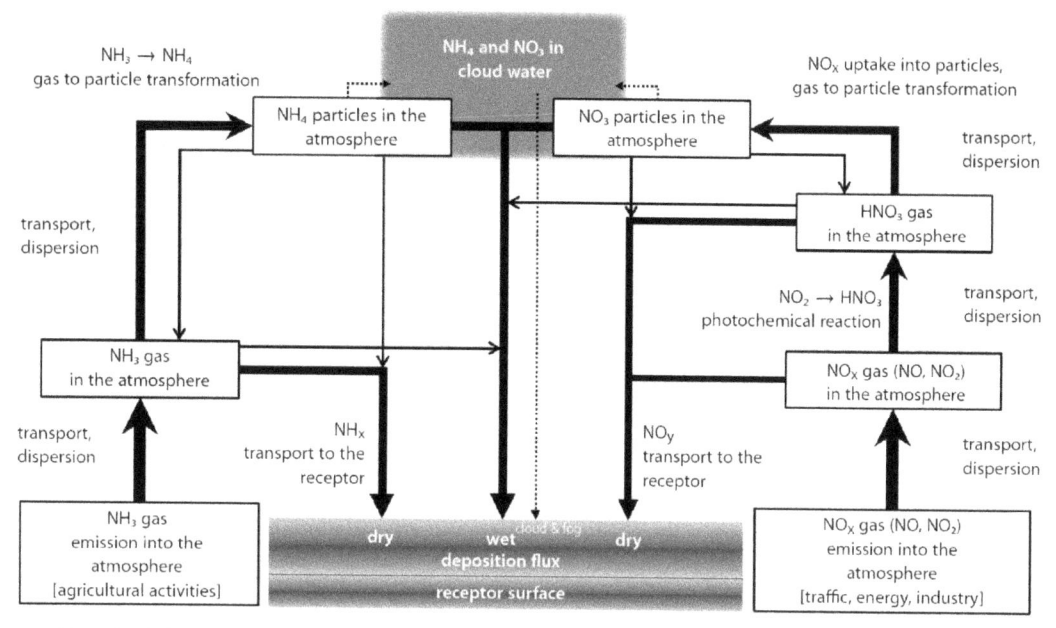

Figure 1.2. Nitrogen emission and deposition pathways

Note: N₂O is excluded as it does not play an important role in nitrogen deposition. For simplicity, reservoir compounds like HONO, HO₂NO₂, peroxyacetylnitrate (PAN) and PAN-like species have also been disregarded.
Source: Adapted by Th. Gauger from Hertel et al. (2006).

1.4.2 Greenhouse balance and ozone layer

With respect to the risk of a greenhouse effect, N_2O, CO_2 and CH_4 remain in the atmosphere long enough to become well mixed, meaning that the amount that is measured in the atmosphere is roughly the same all over the world, regardless of the source of the emissions. Dinitrogen and N_2O are the two most important end-products of the nitrogen cycle, dinitrogen usually being produced in excess of N_2O. Following Portmann et al., 2012, N_2O is relatively inert in the troposphere. It is then transported to the stratosphere where most (90%) is eventually broken down through interaction with high-energy light (i.e. photolysis) into dinitrogen passing through different oxidised nitrogen forms. N_2O is a stock pollutant in the troposphere. This is because N_2O is long-lived – with a tropospheric lifetime of over a century – and as only 7% of N_2O returns to the land surface by deposition.[20]

N_2O provides a natural source and is the main supplier of NO_x to the stratosphere through oxidation to NO (Figure 1.3).[21] The residence time of NO_x in the stratosphere is 1-2 years. In the stratosphere NO reacts rapidly with O_3 to produce NO_2, which releases O_3 by photolysis ("null cycle") or reacts with O, resulting in a net loss of O_3 ("loss cycle").[22]

Figure 1.3. Sources of stratospheric nitrogen oxides (NO$_x$)

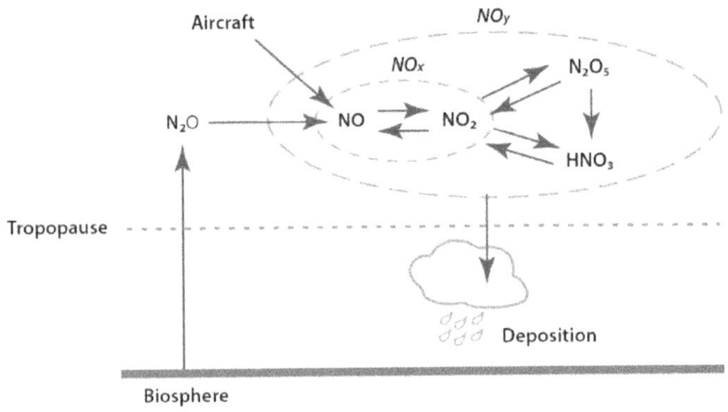

Notes:
NO$_y$ is the sum of all oxidised nitrogen forms.
Nitric oxide (NO) is also produced by aircrafts (by oxidation of dinitrogen at the high temperatures of the aircraft engine)
The tropopause is the boundary between the troposphere and the stratosphere.
Source: Jacob (1999).

1.4.3 Water

Nitrogen pathways in freshwater include dinitrogen fixation, denitrification and nitrogen deposition, exports to sea, and exchange with the sediments (Figure 1.4). In addition to direct discharges by industry and wastewater treatment plants, nitrogen finds its way into the surface waters via pathways that are mainly fed by agriculture such as groundwater, drainage water and runoff. The NO$_3^-$ inputs into groundwater (i.e. NO$_3^-$ not used by crops or denitrified by soil bacteria) are usually the result of leaching from the soil (Durand et al., 2011), where the leaching potential is a function of soil type, crop, climate, tillage practices, fertiliser management, and irrigation and drainage management (USEPA, 2003). The risk of leaching is higher for coarse textured soils and for crops with poor efficiency of nitrogen use (ibid). Nitrogen exports from surface water to downstream estuaries (via riverine and groundwater flow) can be expected to be watershed specific, being typically lower in less developed, highly forested watersheds than in highly developed urban watersheds. Atmospheric depositions of nitrogen into lakes, wetlands and marine waters add to the amounts carried in from rivers (Swackhamer, 2004).

Figure 1.4. Nutrient pathways in freshwater

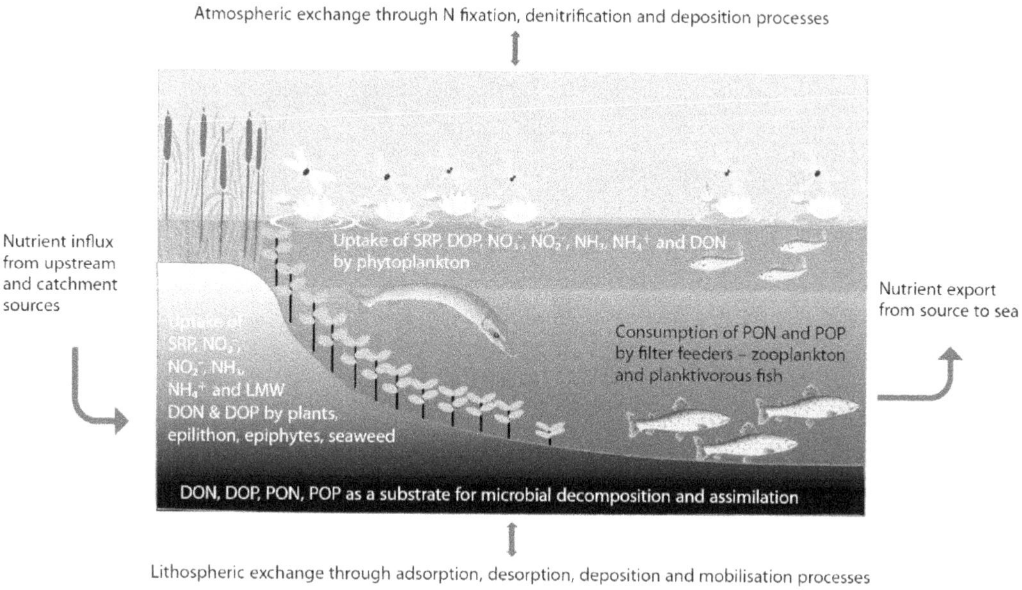

DON, PON = Dissolved Organic Nitrogen; Particulate Organic Nitrogen
DOP, POP = Dissolved Organic Phosphorus; Particulate Organic Phosphorus
SRP = Soluble Reactive Phosphorus
LMW: low molecular weight
Source: Johnes (2016).

Nitrogen pathways in marine water include nitrogen imports by river discharge and precipitation; dinitrogen fixation by cyanobacteria, bacterial remineralisation of dead particulate biomass in sediments and denitrification. The nitrogen budget of continental shelf systems is primarily governed by exchanges with the open ocean (Figure 1.5).

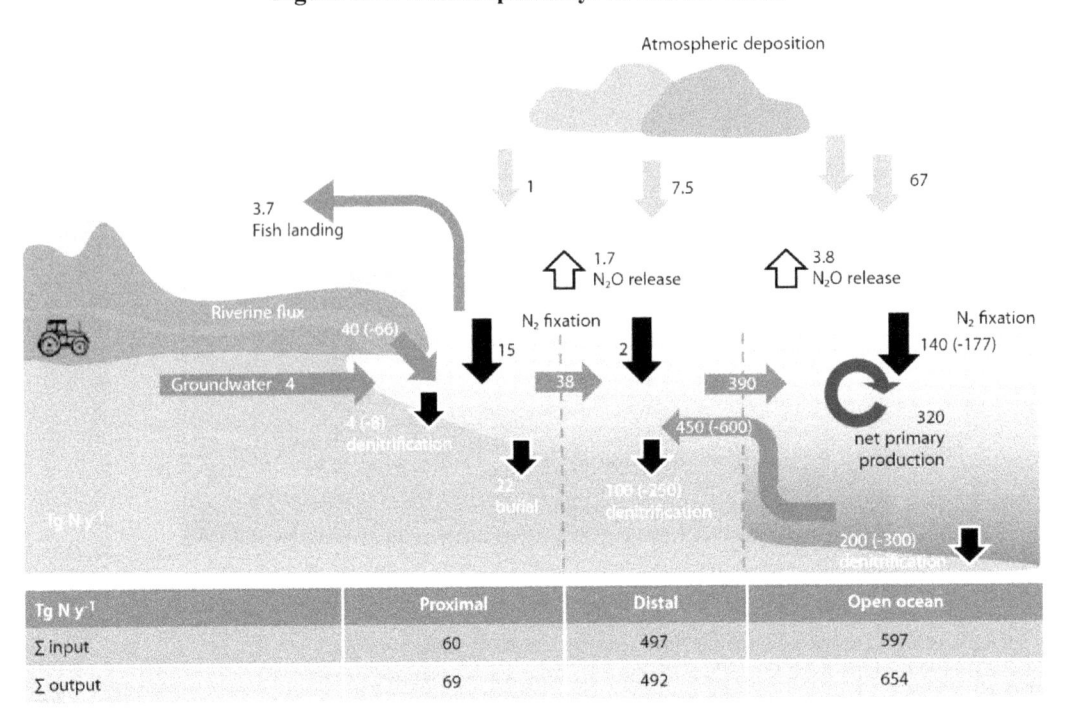

Figure 1.5. Nutrient pathways in marine water

Tg N y⁻¹	Proximal	Distal	Open ocean
Σ input	60	497	597
Σ output	69	492	654

Source: Voss et al. (2013).

Oceanic currents play a role in the exchange of nitrogen within the sea and the resuspension of sediment nitrogen. They may shift impacts long distances away from nutrient sources. For example, recent scientific evidence – prompted by impacts on the tourism industry – revealed that nutrients originating from the Amazonian river basin, where soils are washed away by the rains as a result of deforestation and intensive agriculture, were a major contributor to algal blooms in Caribbean coastal areas.[23] Nutrient-rich waters can also replace nutrient-poor waters in some coastal areas of the ocean and large lakes (such as the North American Great Lakes). Indeed, there is mass continuity in the ocean and large lakes (water being a continuous fluid), so that a change in distribution of water in one area is accompanied by a compensating change in water distribution in another area.

1.4.4 Ecosystems and biodiversity

Apart from acid rain, little recognition had been given to the environmental consequences of nutrients (and toxic substances) that fall from the air as wet and dry deposition onto land-based and aquatic ecosystems (Swackhamer et al, 2004). Yet, the deposition of NH_4^+ and NO_3^- on terrestrial ecosystems favours some species over others (see Annex A). It leads to the accumulation of organic nitrogen in the soil and there is strong evidence that ecological communities respond to the accumulated pool of plant-available nitrogen in the soil, even if the long-term implications are not clear. It is very likely that nitrogen deposition acts synergistically with other stressors, in particular climate change, acid deposition, and GLO; these synergies are poorly understood. The nature and rate of recovery of biodiversity from nitrogen pollution is not well understood. The point is that even if nitrogen deposition rates were to be significantly reduced in the future, habitat recovery would be slow (Plantlife and Plant Link UK, 2017).

Acid rain is a cumulative problem for non-alkaline soils; as the acid-neutralising capacity of soils gets depleted, the ecosystems become increasingly sensitive to additional acid inputs. In contrast, acid rain falling over the oceans is rapidly neutralised by the large supply of carbonate ions. Acid rain falling over regions with alkaline soils or rocks is also quickly neutralised once the acid has deposited to the surface.

1.4.5 Soil

Most of the transformations involved in the biogeochemical nitrogen cycle are carried out by microorganisms in water and soil (in part, following Söderlund and Rosswall, 1982). In non-alkaline soils, NH_3 tends to converts into NH_4^+. Under aerobic conditions (in the presence of oxygen), certain autotrophic bacteria – nitrifying bacteria – can use NH_4^+ oxidation as an energy-yielding process. They can oxidise NH_4^+ to NO_2^- and NO_3^-.[24] Because it releases hydrogen, nitrification also contributes to the acidification of soils.[25]

Under anaerobic conditions (in the absence of oxygen), certain aerobic bacteria – denitrifying bacteria – can use NO_3^- and NO_2^- in place of oxygen, reducing it to dinitrogen. In the process, denitrification generates N_2O and NO.[26]

Plant-soil interactions are mainly governed by interactions between the carbon and nitrogen cycles (in part, following Müller and Clough, 2014). Microbial activity in soils is driven by "rhizodeposition"[27]. As such, nitrogen cycling is closely associated with plant productivity. Conditions that favour plant carbon assimilation may also enhance rhizodeposition. Conversely, a higher demand for nitrogen by plants (e.g. favourable growth temperatures) increases competition for nitrogen between plants and microbes, potentially affecting microbial life. However, very few studies have examined the links between the nitrogen cycle, plant activity, and associated changes in microbial diversity.

Plants utilise mineral nitrogen (NH_4^+ or NO_3^-) and organic nitrogen that enters the soil either via rhizodeposition or microbial mineralisation (DON). DON uptake is reported to be more significant under conditions of nitrogen limitation and low pH. Despite the significance of DON in agricultural ecosystems, current knowledge of the soil DON dynamics is still limited. The role DON plays in gaseous nitrogen loss pathways is also under-researched. Evaluation of the various simultaneously occurring nitrogen transformations in plant–soil systems that lead to mineralisation and immobilisation of nitrogen in the rhizosphere is lacking.

Nitrogen may be lost from soils as NO_3^-, via leaching, or in gaseous forms such as NH_3, NO, N_2O or dinitrogen (Figure 1.6). Denitrification is a key pathway in the soil nitrogen cycle, but a pathway that is still poorly understood. For example, there is uncertainty about soil NO emissions, which will likely increase with increasing anthropogenic sources of nitrogen but with much variability. Research is also needed to better understand the importance of "dissimilatory NO_3^- reduction to NH_4^+" or DNRA. While the available organic carbon and oxygen have a major impact on total denitrification, the soil pH mainly influences the

N₂O/dinitrogen ratio (the ratio tends to increase with decreasing pH). However, there is a lack of field methods to quantify the N₂O/dinitrogen ratio in situ.

Figure 1.6. Nitrogen pathways in soil

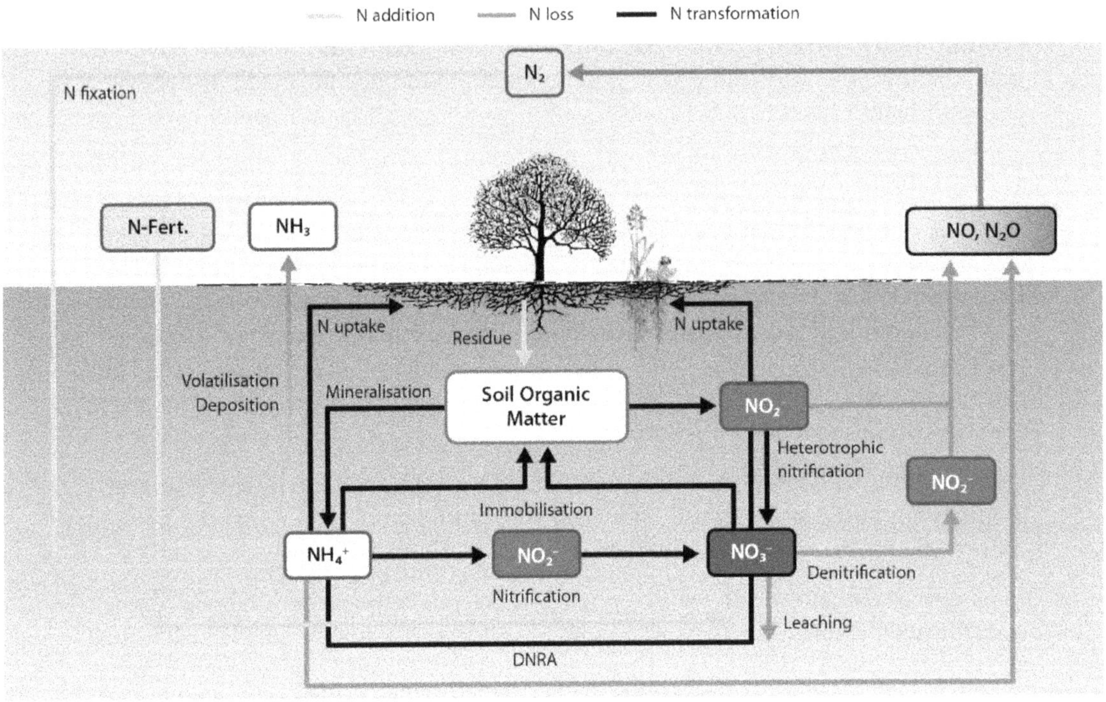

N-Fert = nitrogen fertiliser
DNRA= dissimilatory nitrate (NO₃⁻) reduction to ammonium (NH₄⁺)
Source: Müller and Clough (2014).

The N₂O/dinitrogen ratio of denitrification is most likely governed by the existing microbial communities. In turn soil microbial communities are influenced by environmental conditions (oxygen, temperature, pH), the availability of carbon substrates and the potential for reduction–oxidation reactions.[28] But the interactions between these factors are not fully understood. In fertilised soils, carbon rather than NO₃⁻ availability limits total denitrification. The carbon substrate determines the efficiency with which nitrogen oxides (NO₃⁻, NO₂⁻) are reduced.

1.5   ...and multiple impacts

1.5.1 Air quality

In the atmosphere, NO₂ is directly harmful to human health (it increases likelihood of respiratory problems) (see Annex A). Nitrogen also contributes to the most serious air pollution problems for human health, namely airborne PM and, to a lesser extent, GLO (OECD, 2012). Nitrogen is a precursor of both: NH₃/NH₄⁺ and NOₓ for secondary PM;[29] NOₓ for GLO.[30] The effects of PM can range from eye and respiratory irritation to cardiovascular disease, lung cancer[31] and consequent premature death. The PM of most concern are small (PM₁₀) and especially fine (PM₂.₅) particles, as these are small enough to be able to penetrate deeply into the lungs. Globally, 8% of lung cancer deaths, 5% of cardiopulmonary deaths and around 3% of respiratory infection deaths can be attributed to exposure to $PM_{2.5}$ alone (WHO, 2009). Huang et al., 2017 estimated that NO_3^- and NH_4^+ aerosols accounted for 30% of $PM_{2.5}$ emissions measured in Beijing from June 2014 to April 2015. Exposure to high levels of GLO increases the risk of premature death from lung disease; GLO also affects vegetation by damaging leaves and reducing growth: exposure during the growing season, including at low levels, can have ecosystem-wide impacts as well as economic costs for cultivated land.[32]

1.5.2 Greenhouse balance and ozone layer

As a greenhouse gas (GHG), N_2O produces a positive climate forcing, or global warming effect. N_2O is more effective at warming Earth than CO_2. The two most important characteristics of a GHG in terms of climate impact are how well the gas absorbs energy from light (either directly or after reflection from the earth's surface), and how long the gas stays in the atmosphere.[33] The Global Warming Potential (GWP) is a measure of the total energy that a unit mass of a gas absorbs over a particular period of time (usually a 100 year period is used). The larger the GWP, the more warming the gas causes. CO_2 has a GWP of 1 and serves as a baseline for other GWP values. N_2O is a more powerful GHG than CO_2 and, with a GWP 265 times higher over a 100-year time scale, is responsible for about 6% of the worldwide anthropogenic greenhouse effect. However, the GWP metric is not a perfect measure of climatic impacts. N_2O persists in the atmosphere for about a century, compared to thousands of years for CO_2, whereas the average lifetime of CH_4 is around 12 years (Allen et al., 2016). For long-lived GHGs, it is the cumulative GHG emissions that largely determine the extent of future climate change, not just the emissions rate in a given year. So the continually rising concentration of N_2O in the atmosphere is of concern in relation to the stringent climate goals in the Paris Agreement.

Nitrogen-containing aerosols have an offsetting greenhouse effect (see Annex A). However, N_2O remains much longer in the atmosphere than aerosols (over 120 years compared to weeks or months), so if such aerosols are reduced but N_2O remains at the same concentration or increases, the net effect would change from marginally negative to more strongly positive. The N_2O greenhouse effect is also partially offset by increased CO_2 uptake in terrestrial ecosystems due to atmospheric nitrogen deposition (see Annex A).

N_2O also contributes significantly to the depletion of the stratospheric ozone layer (O_3) that protects life on Earth by absorbing some of the ultraviolet rays from the sun. Indeed, N_2O provides a natural source of NO_x to the stratosphere and, with the phasing-out of chlorofluorocarbons (CFCs), NO_x has become a major depleting threat for O_3 in the stratosphere (Ravishankara et al., 2009) and is currently unregulated by the Montreal Protocol on Substances that Deplete the Ozone Layer (a protocol to the Vienna Convention for the Protection of the Ozone Layer).

1.5.3 Water quality

In water, nitrate (NO_3^-) is directly harmful for human health (high concentrations in drinking water can cause blood disorder in infants) (see Annex A). NO_3^- in drinking water may also increase risk of colorectal cancer due to endogenous transformation into carcinogenic N-nitrosocompounds (Jörg Schullehner et al., 2018).

High concentrations of nutrients (i.e. NO_3^- and phosphorus) in fresh and marine waters result in phytoplankton (microscopic algae) growth, a process called "eutrophication". A high density of phytoplankton reduces the water transparency and the reduced penetration of sunlight limits the depth to which macrophytes and sea grasses can grow. One of the most striking effects of excessive levels of nutrients is the formation of low oxygen (hypoxic) or oxygen-free (anoxic) zones in deep water layers where the higher organisms are unable to survive, the so-called "dead zones" (Breitburg et al., 2018; see also Chapter 3). A distinction must be made between marine waters and fresh waters. Phosphorus is often the driver of eutrophication in freshwater while it is NO_3^- in marine waters.

1.5.4 Ecosystems and biodiversity

Excess nitrogen from air-based and land-based sources is one of the major drivers of biodiversity loss in Europe (Sutton et al., 2011). Nitrogen impacts vegetation diversity through direct foliar damage, eutrophication, acidification, and susceptibility to secondary stress. The accumulation of extra nutrients, as well as reduction in soil pH, is negatively affecting natural and semi-natural habitats whose important biodiversity developed in direct response to low nutrient levels (Plantlife and Plant Link UK, 2017). Nitrogen deposition therefore contributes to biodiversity loss (nitrogen-loving species outcompete other species). While there is a wealth of evidence on the magnitude, components and effects of nitrogen deposition on floral biodiversity in Europe and North America, there is an obvious lack of information on impacts on above- and below-ground fauna, and all impacts in other parts of the world, with no clear overview of how the different strands of evidence fit together (Sutton et al., 2014).

It is not yet clear if different wet-deposited forms of nitrogen (e.g. NO_3^- versus NH_4^+) have different effects on biodiversity. However, NH_3 can be particularly harmful to vegetation, especially lower plants, through direct foliar damage (Sutton et al., 2011). For example, a long-term controlled field experiment showed that, for the same overall nitrogen load, NH_3 deposition was more damaging to bog vegetation than deposition of ammonium aerosol, which was more damaging than deposition of nitrate aerosol (Phoenix et al., 2012). However, responses to the form of nitrogen are complex and habitat dependent, with conversion between nitrogen forms resulting from the activities of soil microbes (Plantlife and Plant Link UK, 2017).

The deposition of NO_2, NH_3 and sulphur dioxide (SO_2) acidifies terrestrial and freshwater ecosystems ("acid rain").[34] Acid rain remains an issue (including in the OECD area) mainly because the decrease in NO_x emissions has not been commensurate with the decrease of SO_2 emissions. Eutrophication is a consequence of excess input of nitrogen nutrients (NH_3, NO_x); the atmospheric input of other nutrients is negligible.

Acid rain has negative effects on freshwater ecosystems. Elevated acidity in a lake or river is directly harmful to fish because it corrodes the organic gill material and attacks the calcium carbonate skeleton. In addition, the acidity dissolves toxic metals such as aluminium from the sediments. Acid rain is also harmful to terrestrial vegetation with little acid-neutralising capacity,[35] mostly because it leaches nutrients such as potassium and allows them to exit the ecosystem by runoff. Beyond the input of acidity, the deposition of NH_4^+ and NO_3^- is an important contributor in the eutrophication of aquatic ecosystems (by providing ecosystems with a source of directly assimilable nitrogen in addition to land-based sources). Acidifying deposition may also damage building structures and monuments.

1.5.5 Soil quality

The major nitrogen threats on soil quality for both agricultural and natural soils are related to changes in soil acidification, and loss of soil diversity (Sutton et al., 2011). Soil acidification may lead to a decrease in crop and forest growth and leaching of components negatively affecting water quality, including heavy metals.

The effect of nitrogen on soil organic matter content is uncertain. The effect of nitrogen on diversity of soil (micro) organisms and the effects of changes of soil biodiversity on soil fertility, crop production and nitrogen emissions towards the environment are not fully understood.

1.6 The "nitrogen cascade"

In addition to the doubling of production since preindustrial time, nitrogen has another feature that distinguishes it from other pollutants - it can go a long way once released into the environment. Once created, the same nitrogen atom can cause multiple effects in the atmosphere, in terrestrial ecosystems, in freshwater and marine systems, and on the climate, as it moves through the biogeochemical pathways. Galloway et al., 2003 call this sequence of effects the "nitrogen cascade" (Figure 1.7).

Figure 1.7. The nitrogen cascade

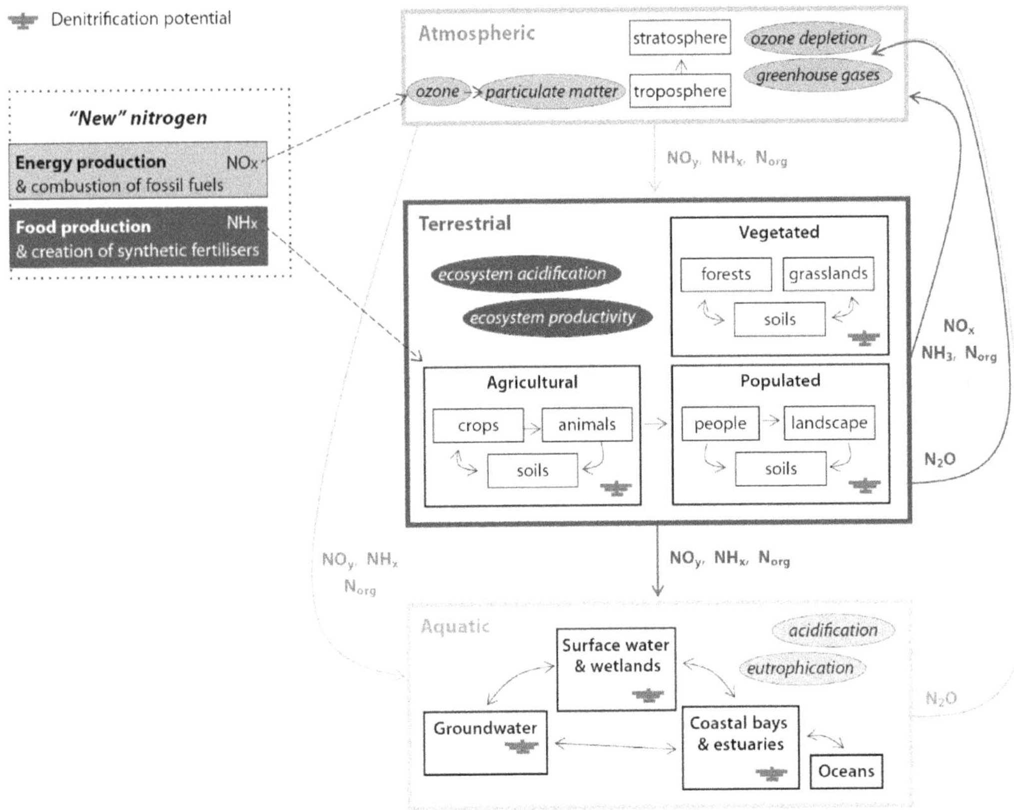

Note: Conceptual diagram illustrating the cascade of nitrogen from point of origin along its biogeochemical pathways, with the associated negative impacts.
Source: USEPA Science Advisory Board (undated), nrcs.usda.gov/Internet/FSE_DOCUMENTS/nrcs143_008785.pdf.

For example, the biogeochemical journey of an atom of nitrogen from its point of formation could be as follows. NO_x formation during fossil fuel combustion first has the potential to contribute to the creation of GLO, a component of smog,[36] then can be converted to HNO_3 which is a major component of acid deposition, or in the atmosphere it can be converted to an aerosol which will decrease light scattering[37] and promote formation of cloud drops. Once removed from the atmosphere, HNO_3 and nitrate aerosols can cause both fertilisation and acidification of soils which in turn (via NO_3^- leaching) will result in acidification of low alkalinity freshwaters and fertilisation of those same waters. Upon transport to coastal regions and lakes, the same nitrogen atom (as it cascades) can contribute to eutrophication with resulting loss of freshwater and marine biodiversity. As a final step, the nitrogen atom in soil and water can be converted to N_2O, which contributes to the greenhouse effect in the troposphere and ozone destruction in the stratosphere. Figure 1.8 illustrates another series of possible cascading nitrogen effects from one environmental media to another.

Figure 1.8. Example of serial impacts on health and the environment of a single nitrogen atom

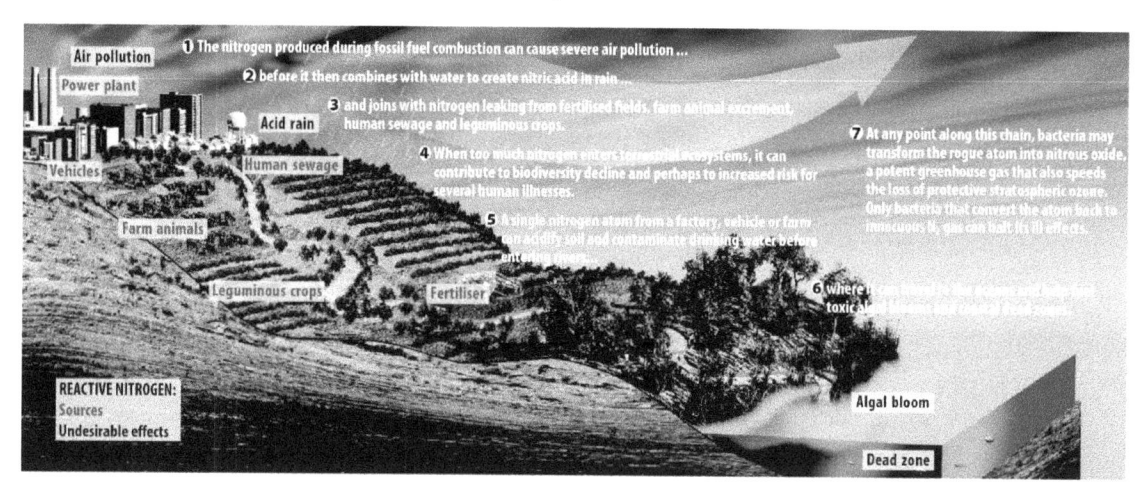

Source: Townsend and Howarth (2010).

Notes

[1] "Nitrogen" refers to the reactive forms of nitrogen (as opposed to "dinitrogen" – N_2 - which refers to the inert form).

[2] Dinitrogen accounts for 78% of air on a molar (i.e. number of molecules) basis (Holloway and Wayne, 2015).

[3] Until one century ago, Chilean saltpetre (and before it Peruvian guano) was the main source of nitrogen for world agriculture and industry.

[4] Named after its inventors, the German chemists Fritz Haber and Carl Bosch. The process converts dinitrogen to ammonia (NH_3) by a reaction with hydrogen (H_2) -- $N_2 + 3 H_2 \rightarrow 2 NH_3$ -- using a metal catalyst under high temperatures and pressures.

[5] This is consistent with the estimates of Vitousek et al., 2013 and Erisman and Larsen, 2013, for which current human nitrogen production is about the same as total natural production (terrestrial and oceanic).

[6] Percentages are rounded figures based on data in Table 1.1. They are broadly consistent with Sutton et al, 2011 who estimated that during the 2000s, food production (use of fertilisers, manure and leguminous crops) accounted for three quarters of the nitrogen produced by humans, the combustion of fossil fuels and industrial processes accounting equally for the remainder.

[7] Jacob, 1999 estimates that natural sources generate 28% of total NO_x emissions. This includes microbial denitrification in soils (13%), oxidation of NH_3 emitted by the biosphere (7%), lightning (7%) and transport from the stratosphere (1%).

[8] Nitrogen-enriched fuel, as paradoxical as it may seem, can reduce NO_x emissions per kilometre driven. It can slightly increase NO_x emissions per litre of fuel burned, but it helps to increase fuel efficiency because nitrogen breaks down carbon deposits on moving parts of the engine (less fuel is burned).

[9] NO_x is produced when dinitrogen reacts with oxygen under high temperature and pressure conditions in the combustion chamber of the vehicle. However, understanding the chemical processes that govern the formation and destruction of NO_x during combustion continues to be a challenge (Glarborg et al., 2018).

[10] Other sources of NH_3 comprise industries, landfills, household products, biomass burning, motor vehicles, and manure from wild animals.

[11] These include nitrate radicals (NO_3), nitrous acid (HONO), nitric acid (HNO_3), dinitrogen pentoxide (N_2O_5), nitryl chloride ($ClNO_2$), peroxynitric acid (HNO_4), peroxyacetyl nitrate and its homologues (PANs), other organic nitrates, such as alkyl nitrates (including isoprene nitrates).

[12] HNO_3 is produced by atmospheric oxidation of NO_x in the daytime and via O_3 at night.

[13] An aerosol is a suspension of a solid or liquid PM in the air.

[14] Globally, more than 80% of GLO results from chemical production within the troposphere and less than 20% are supplied to the troposphere by transport from the stratosphere.

[15] For example, isoprene emissions from natural vegetation (isoprene is a by-product of photosynthesis).

[16] Perhaps because phosphorus limitation prevents aquatic life from developing.

[17] N₂O is a relatively inert gas and thus is able to float through the troposphere without being destroyed by other atmospheric gases.

[18] HNO₃ deposition (acid rain) may also affect buildings (by dissolving the calcite in marble and limestone).

[19] Most of the atmospheric nitrogen falls into two broad categories: oxidised nitrogen and reduced nitrogen. The oxidised form of nitrogen is mainly dominated by nitrogen oxides (NO$_x$) and the reduced form by ammonia species (NH₃ and NH₄⁺, denoted NH$_x$).

[20] Around 10% is oxidised in the stratosphere to produce NO$_x$, 30% of which is then converted into dinitrogen and 70% deposited to the land surface (Portmann et al., 2012). Thus only 7% of N₂O eventually returns to the land surface by deposition.

[21] In the stratosphere, N₂O encounters high concentrations of O, allowing oxidation to NO.

[22] $NO_2 + h\nu \xrightarrow{O_2} NO + O_3$ (O₃ null cycle) or $NO_2 + O \rightarrow NO + O_2$ (O₃ loss cycle).

[23] The same phenomenon of accumulation of Sargassum (also known as Gulfweed) is occurring off Benin and Sierra Leone, some distance from the mouth of the Congo River.

[24] Since the first description of nitrifying bacteria more than 100 years ago, it was thought that nitrification was achieved by the joint activity of two groups of microorganisms - ammonia- and nitrite-oxidisers. Daims et al., 2015 discovered that complete nitrifiers exist that can oxidise as single microorganisms NH₄⁺ to NO₃⁻.

[25] The adsorption of NH₄⁺ on the clay-humic complex in place of hydrogen causes the acidification of the soil by the hydrogen thus released.

[26] The main steps in denitrification are as follows: NO₃⁻ > NO₂⁻ > NO > N₂O > N₂.

[27] During their life, plant roots release organic carbon and organic nitrogen (e.g. amino acids) into their surrounding environment; this process is named "rhizodeposition".

[28] Or "redox" (i.e. the availability of electron receptors such as nitrogen oxides).

[29] PM can be divided into two types: primary PM emitted directly to the atmosphere, such as black carbon; and secondary PM formed in the atmosphere from precursor gases.

[30] Other precursors of GLO are VOCs, including CH₄, and to a lesser extent, CO.

[31] Toxic and carcinogenic pollutants like heavy metals or polycyclic aromatic hydrocarbon are frequently bound to PM.

[32] GLO is toxic to humans and vegetation because it oxidises biological tissue.

[33] In part, following epa.gov/climateleadership/atmospheric-lifetime-and-global-warming-potential-defined, accessed 2 February 2018.

[34] HNO₃ also contributes to the formation of acid rain by dissociating in rainwater to release H⁺.

[35] For example, areas sensitive to acid rain over North America include New England, eastern Canada, and mountainous regions, which have granitic bedrock and thin soils.

[36] The term "smog" was first used around 1950 to describe the combination of smoke and fog in London. Today, it refers to a mixture of pollutants made up mostly of GLO (https://www.epa.gov/air-pollution-transportation/smog-soot-and-local-air-pollution, accessed 17 March 2018).

[37] And can also reflect some sunlight back into space, offsetting some of the warming due to GHGs.

References

Abrol, Y.P. et al. (2017), *The Indian Nitrogen Assessment, Sources of Reactive Nitrogen, Environmental and Climate Effects, Management Options, and Policies*, Elsevier.

Allen, M.R. et al. (2016), "New Use of Global Warming Potentials to Compare Cumulative and Short-lived Climate Pollutants", *Nature Climate Change*, 6(8).

Breitburg, D. et al. (2018), "Declining Oxygen in the Global Ocean and Coastal Waters", *Science*, 359 (6371).

Daims, H. et al. (2015), "Complete Nitrification by Nitrospira Bacteria", *Nature*, 528.

Denman, K.L.et al. (2007), "Couplings Between Changes in the Climate System and Biogeochemistry", in *Climate Change 2007: The Physical Science Basis*, contribution of Working Group I to the 4th Assessment Report of the Intergovernmental Panel on Climate Change (IPCC), Cambridge University Press.

Durand, P. et al. (2011), "Nitrogen Processes in Aquatic Ecosystems", In: Sutton M.A. et al (Eds.), *The European Nitrogen Assessment. Sources, Effects and Policy Perspectives*, Cambridge University Press.

Erisman, J.W. et al. (2008), "How a Century of Ammonia Synthesis Changed the World", *Nature Geoscience*, 1, October 2008.

Erisman, J.W. and T.A. Larsen (2013), "Nitrogen Economy of the 21st Century", in *Source Separation and Decentralisation for Wastewater Management*, T.A. Larsen et al. (eds), IWA Publishing, London.

Fowler, D. et al. (2013), "The Global Nitrogen Cycle in the Twenty-first Century", *Phil. Trans. R. Soc. B*, 368(1621).

Galloway, J.N. et al. (2003), "The Nitrogen Cascade", *BioScience*, 53 (4).

Gauger, Th. (2018), "Modelling and Mapping Air Concentration and Atmospheric Deposition of Reactive Nitrogen Species in Baden-Württemberg for 2012 to 2016", Institute of Navigation, University of Stuttgart.

Geupel, M. (2015), "Towards a National Nitrogen Strategy for Germany", presentation to the OECD Environment Policy Committee (EPOC), 6-8 October 2015, Federal Environment Agency Germany, Section II 4.3 Air Quality Control and Terrestrial Ecosystems.

Glarborg, P. et al. (2018), "Modeling Nitrogen Chemistry in Combustion", *Progress in Energy and Combustion Science*, 67.

Goldberg, S.J. et al. (2015), « Refractory Dissolved Organic Nitrogen Accumulation in High-elevation Lakes", *Nature Communications*, 6:6347.

Hager, T. (2009), *The Alchemy of Air*, Broadway Books.

Hellsten, S. et al. (2017), *Nordic Nitrogen and Agriculture, Policy, Measures and Recommendations to Reduce Environmental Impact*, Nordic Council of Ministers, TemaNord 2017:547, doi.org/10.6027/TN2017-547.

Hertel, O. et al. (2011), "Nitrogen Processes in the Atmosphere", In: Sutton M.A. et al (Eds.), *The European Nitrogen Assessment. Sources, Effects and Policy Perspectives*, Cambridge University Press.

Hertel, O. et al. (2006), "Modelling Nitrogen Deposition on a Local Scale: a Review of the Current State of the Art", *Environmental Chemistry*, 3(5).

Holloway, A.M. and R.P. Wayne (2015), *Atmospheric Chemistry*, Royal Society of Chemistry.

Houlton, et al. (2018), "Convergent Evidence for Widespread Rock Nitrogen Sources in Earth's Surface Environment", *Science*, 360(6384).

Huang, X. et al. (2017), "Chemical Characterization and Synergetic Source Apportionment of $PM_{2.5}$ at Multiple Sites in the Beijing–Tianjin–Hebei Region, China", *Atmos. Chem. Phys. Discuss.*, 17.

Jacob, D.J. (1999), *Introduction to Atmospheric Chemistry*, Princeton University Press.

Johnes, P. (2016), "Nitrogen Pollution of Inland and Coastal Waters: Sources, Impacts and Opportunities", presentation to the Joint OECD/TFRN Workshop on *The Nitrogen Cascade and Policy – Towards Integrated Solutions*, OECD, Paris, 9-10 May 2016.

Lefohn, A.S. et al. (2014), « Estimates of Background Surface Ozone Concentrations in the United States Based on Model-derived Source Apportionment", *Atmospheric Environment*, 84.

Müller, C and T. J. Clough (2014), "Advances in Understanding Nitrogen Flows and Transformations: Gaps and Research Pathways", *Journal of Agricultural Science*, Special Issue from the 17th International Nitrogen Workshop, 152 (S1).

OECD (2012), *OECD Environmental Outlook to 2050: The Consequences of Inaction*, OECD Publishing, Paris, doi.org/10.1787/9789264122246-en.

Phoenix, G.K. et al. (2012), "Impacts of Atmospheric Nitrogen Deposition: Responses of Multiple Plant and Soil Parameters Across Contrasting Ecosystems in Long-term Field Experiments", *Global Change Biology*, 18.

Plantlife and Plant Link UK (2017), "We Need to Talk About Nitrogen, The Impact of Atmospheric Nitrogen Deposition on the UK's Wild Flora and Fungi", January 2017.

Portmann, R.W. et al. (2012), "Stratospheric Ozone Depletion due to Nitrous Oxide: Influences of Other Gases", *Philos Trans R Soc Lond B Biol Sci.*, 367(1593).

Prentice, I.C. et al. (2015), "Biosphere Feedbacks and Climate Change", Grantham Institute, Briefing Paper 12, Imperial College London.

Ravishankara, A.R. et al. (2009), « Nitrous Oxide (N_2O): the Dominant Ozone-depleting Substance Emitted in the 21st Century", *Science*, 326(5949).

Richmond-Bryant, J. et al. (2016), "Estimation of On-road NO_2 Concentrations, NO_2/NO_x Ratios, and Related Roadway Gradients from Near-road Monitoring Data", submitted to *Air Quality, Atm and Health*.

Schullehner, J. et al. (2018), « Nitrate in Drinking Water and Colorectal Cancer Risk: A Nationwide Population-based Cohort Study", *International Journal of Cancer*, 23 February (Epub ahead of print).

Seitzinger, S.P. et al. (2010), "Global River Nutrient Export: A Scenario Analysis of Past and Future Trends", *Global Biogeochemical Cycles*, 24(4), doi.org/10.1029/2009GB003587.

Söderlund, R. and T. Rosswall (1982), "The Nitrogen Cycles", in Hutzinger, O. (edit), *The Handbook of Environmental Chemistry*, Vol. 1 Part B (The Natural Environment and the Biogeochemical Cycles), Springler-Verlag, Berlin Heidelberg GmbH.

SRU (2015), *Nitrogen: Strategies for Resolving an Urgent Environmental Problem*, German Advisory Council on the Environment, Berlin.

Stocker, B.D. et al. (2016), "Terrestrial Nitrogen Cycling in Earth System Models Revisited", *New Phytologist*, 210(4), doi.org/10.1111/nph.13997.

Sutton, M.A. et al. (2011), *The European Nitrogen Assessment: Sources, Effects and Policy Perspectives*, Cambridge University Press.

Sutton, M. et al. (2014), *Nitrogen Deposition, Critical Loads and Biodiversity*, Springer.

Swackhamer, D.L. et al. (2004), "Impacts of Atmospheric Pollutants on Aquatic Ecosystems", *Issues in Ecology* (12).

The Economist (2011), "The Anthropocene, a Man-made World", *The Economist*, May 26th 2011.

Tomich, Th. P. et al. (2016), *The California Nitrogen Assessment: Challenges and Solutions for People, Agriculture, and the Environment*, University of California Press.

Townsend, A.R. and R. W. Howarth (2010), "Fixing the Global Nitrogen Problem", *Scientific American*, February 2010.

USEPA (2003), *National Management Measures to Control Nonpoint Source Pollution from Agriculture*, U.S. Environmental Protection Agency, Office of Water, EPA 841-B-03-004.

USEPA-SAB (2011), *Reactive Nitrogen in the United States: An Analysis of Inputs, Flows, Consequences and Management Options*, U.S. Environmental Protection Agency's Science Advisory Board, EPA-SAB-11-013, USEPA, Washington D.C., yosemite.epa.gov/sab/sabproduct.nsf/WebBOARD/INCFullReport/$File/Final%20INC%20Report_8_19_11(without%20signatures).pdf.

Vitousek, P.M et al. (2013), "Biological Nitrogen Fixation: Rates, Patterns and Ecological Controls in Terrestrial Ecosystems", *Phil Trans R Soc B*, 368.

Voss, M. et al. (2013), "The Marine Nitrogen Cycle: Recent Discoveries, Uncertainties and the Potential Relevance of Climate Change", *Phil. Trans. R. Soc. B*, 368.

WHO (2009), *Global Health Risks: Mortality and Burden of Disease Attributable to Selected Major Risks*, World Health Organization, Geneva, who.int/healthinfo/global_burden_disease/GlobalHealthRisks_report_full.pdf.

Chapter 2. Proposed approach to address nitrogen pollution

This chapter proposes a three-pronged approach to cost-effectively respond to nitrogen pollution. First, to better manage the risks of air, soil and water pollution and associated ecosystems through a detailed analysis of the nitrogen pathways, the so-called "spatially targeted risk approach". Second, address the steady increase in nitrous oxide concentrations in the atmosphere through a "global risk approach". Third, take into account the uncertainty of cascading effects and anticipate potentially significant long-term impacts through a "precautionary approach".

As we have seen in Chapter 1., there are specific nitrogen pathways for each environmental medium and the relative impacts on water quality, air quality, greenhouse balance and ozone layer, ecosystems and biodiversity, and soil quality ("WAGES"). The nitrogen cascade described in Chapter 1 "superimposes" these media-specific pathways by including the possibility for a nitrogen atom to pass from a medium to the other, and from one impact to another. How should environmental policy handle this?

- Should environmental policy stick to media-specific impacts,[1] as it already does when managing environmental risks?

- Should environmental policy add a more "precautionary" dimension that takes into account the potential for and uncertainty of cascading effects across media and anticipates potentially large impacts in the longer term?

There is only one form of nitrogen that has a global impact and that is nitrous oxide (N_2O), which impacts on both global warming and the stratospheric ozone layer. While the impacts of accumulating N_2O will differ regionally, the problem can only be tackled globally since it is a long-lived and well-mixed gas in the atmosphere. The other forms of nitrogen are labile (i.e. they move, change form and combine easily with other pollutants); their impacts on air, soil and water quality will therefore tend to be more local or regional. Policies to tackle these different scales of impacts are therefore needed. As we will see in this section, N_2O management calls for global action because of the global coverage of its impacts whereas a more spatially targeted approach is more appropriate for other nitrogen risks.

On the other hand, the uncertainties of the cascade call for a complementary action (in part, following OECD, 2016). However, a cost-effective environmental policy requires prioritising the management of well-documented risks over the management of such uncertainties (Table 2.1). First, there is a need to manage risks by better understanding the pathways of nitrogen between sources and impacts, including the contributions of selected sources at different scales and times. Second, in the absence of strong evidence on pathways, there is a need to prevent nitrogen entering the environment by developing strategies addressed to all sources (on the basis of the most cost effective means) to reduce them. What differs between the two approaches is whether nitrogen pathways leading to specific impacts are sufficiently well understood (or predictable) or are largely uncertain.

Table 2.1. A three-pronged approach to address nitrogen pollution

Spatially targeted risk approach	Global risk approach	Precautionary approach
Nitrogen forms		
All but N$_2$O	N$_2$O	All
Pathways		
Impact-Pathway Analysis (IPA)[1]	Global exposure	Nitrogen cascade
Focus		
Media-specific (air, soil, water)	Greenhouse effect, ozone layer	"Systemic" (all-media)
Intervention points (scale)		
Risk specific	Global	National (based on monitoring the country's total nitrogen load)
Policy effectiveness		
High (tailored to risk)	High (tailored to risk)	Low (only targets the nitrogen load)
Policy priority		
According to policy objectives	According to policy objectives	When IPA is not possible

1. Impact-Pathway Analysis (IPA) is an evaluation of the pathways that generate an impact (including through modelling) to estimate the expected benefits of possible emissions changes.
Source: OECD Secretariat.

Section 2.1 will set out a proposal for how to implement the spatial management of environmental risks related to forms of nitrogen other than N$_2$O as well as a global approach to management of N$_2$O. Section 2.2 will examine how to approach precautionary management of uncertainty related to the nitrogen cascade.

2.1 The risk approach

2.1.1 The different nitrogen risks

Countries have set acceptable levels of health risks for nitrogen dioxide (NO$_2$) and nitrate (NO$_3^-$) concentrations in air and water, respectively, as well as for concentrations of ground-level ozone (GLO) and particulate matter (PM) (to both of which nitrogen is a precursor). Any emission reduction or any practice to achieve it arises from these acceptable levels of risk.[2]

To address ecosystem risks, critical loads and critical levels have been estimated for the terrestrial ecosystems (forests, wetlands) and are used to regulate nitrogen oxides (NO$_x$), ammonia (NH$_3$) and GLO emissions under the Convention on Long-range Transboundary Air Pollution (LRTAP) and its protocols. For example, in some countries Total Maximum Daily Load (TMDL) is used to calculate the maximum amount of nitrogen allowed to enter aquatic ecosystems (lakes, coastal areas) in which the acceptable level of water quality is not met.

As regards climate change, the Paris Agreement established a level of global warming risk[3] acceptable to the parties and, therefore, indirectly, an acceptable level of climate forcing due to greenhouse gases (GHGs), including N$_2$O. All other things being equal, to stop global temperatures increasing, emissions of long-lived GHGs must eventually fall to zero on a net basis. The Montreal Protocol on Substances that Deplete the Ozone Layer sets an acceptable risk of depletion of the ozone layer, but N_2O is not one of the ozone depleting substances covered by the Protocol at present.

Beyond these regulated risks, the risk of undermining the resilience (of exceeding the coping capacity) of nitrogen sinks such as terrestrial biomass change and marine sediments should also be addressed. For example, if the land sinks were to be saturated, the risk of nitrogen export from watersheds to coastal waters and the associated impacts would be much higher. So far, we do not fully understand the resilience of ecosystems to increased nitrogen loading. Although the data are highly uncertain, the storage of nitrogen in soils and trees[4] appears to represent only a small part of the annual nitrogen input into the land sinks, most of which seems to be denitrified (USEPA-SAB, 2011),[5] thus in part (i.e. the N_2O portion) contributing to global climate and ozone risks. There is an even greater uncertainty about the data when it comes to marine sediments. In particular, the diversity of possible cycling mechanisms have hindered the ability to quantify marine organic matter transformation, degradation and turnover rates (Walker et al., 2016).

Cost-benefit analysis (CBA) defines the risks to be managed as a priority (in part, following OECD 2008). Nitrogen risk management can make an important contribution to social welfare (for example by protecting the natural bases of production and improving human health). However, it can also entail significant economic costs. It is therefore important to carefully consider whether the additional benefits of environmental improvements, and the additional costs to society of achieving these improvements, are justified.

2.1.2 Deepening pathway analysis to better manage risks of nitrogen pollution

Current nitrogen management policies have often focused on a specific impact without consideration of biogeochemical pathways[6] contributing to this impact (USEPA-SAB, 2011). Yet, a better knowledge of pathways can improve the cost-effectiveness of risk management by better identifying the points of policy intervention (i.e. by better allocating emission reduction across different nitrogen sources). For example, the reduction of nitrogen impacts in estuaries may greatly benefit from a stricter control of atmospheric deposition in the airshed[7] as well as stricter runoff controls in the watershed. This is why significant efforts have been made to join-up air and water management in the Chesapeake Bay region (Linker et al., 2013).

The Impact-Pathway Analysis (IPA) is an evaluation of the pathways that generate an impact (including through modelling) to estimate the expected benefits of possible emissions changes (in part, following OECD, 2018). IPA differs from damage assessment, which assesses an impact at a given point in time, without explicitly considering how that impact was generated (ibid). IPA recognises that nitrogen can move between environmental media (air, water, soil and biota) as it travels along a pathway from one or more sources to a receptor.

This section discusses the importance of deepening the analysis of pathways to adapt risk management to the specificities of nitrogen pollution. Chapter 3. presents case studies of IPA and its translation into policy-making.

A four-step approach is proposed to implement the IPA:

- First, identify the nitrogen sources of relevance to the impact and delineate the different nitrogen emission zones that converge towards the risk area

- Second, calculate the marginal abatement costs for reductions in release of the different nitrogen forms, for that, estimate the potential for new or additional emission reductions in each emission zone.

- Third, compare the cost-effectiveness of emission reductions for all sources of risk in the different emission zones

- Fourth, estimate the marginal ancillary benefits of reducing nitrogen emissions in the different emission zones (i.e. the avoided damages along the pathways of nitrogen to the risk area).

The first step is to delineate the different nitrogen emission zones that converge towards the risk area (Figure 2.1). The advance of knowledge on pathways will allow consideration of new nitrogen sources (e.g. located further back along the pathways) compared with the absence of IPA, where the only points of intervention to manage the risk of pollution are the sources whose link with the impact is already well established.

Figure 2.1. Nitrogen emission zones for different pollution risks

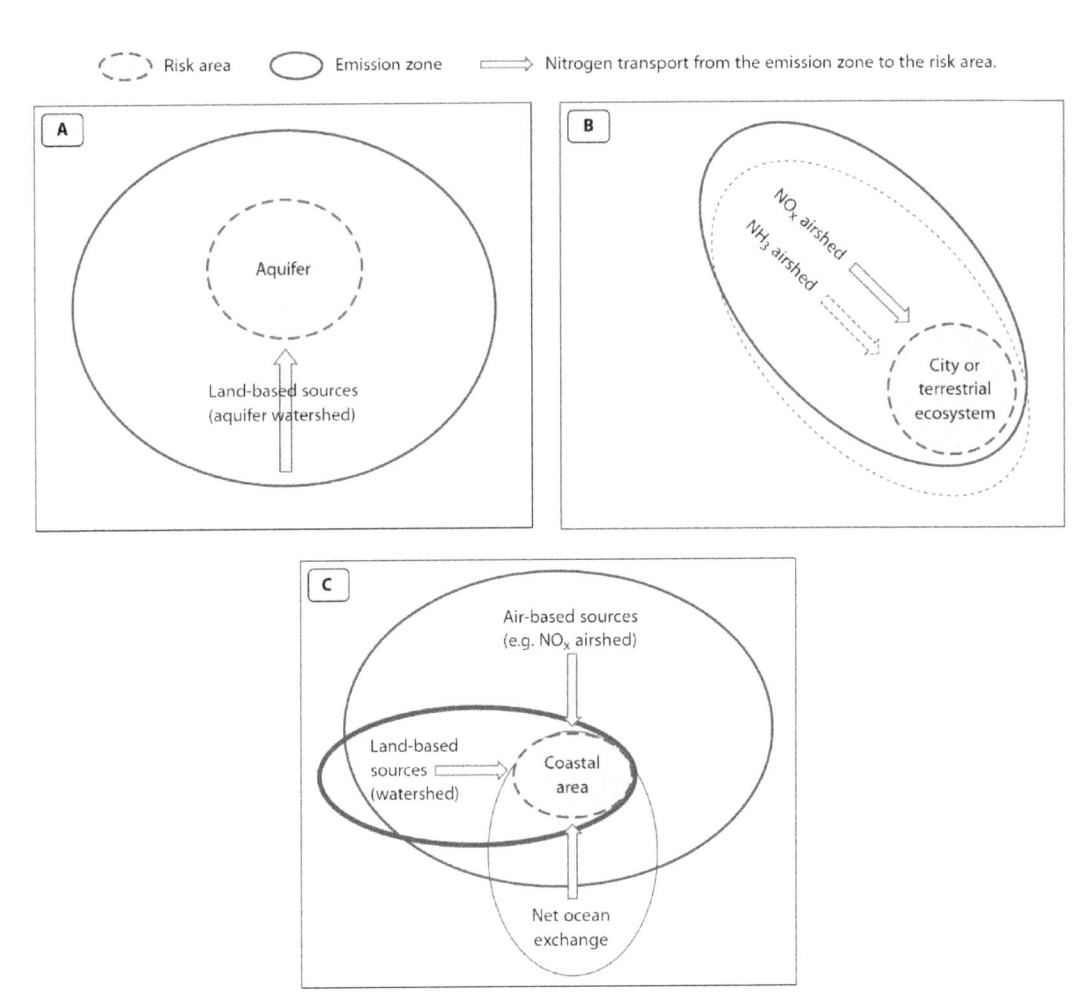

The three panels represent three types of risk:
(A) contamination of an aquifer by nitrate (NO_3^-)
(B) atmospheric pollution of a city or terrestrial ecosystem
(C) nitrogen pollution of a coastal area.
Source: OECD Secretariat.

The Puget Sound dissolved oxygen study is a good example of the delineation of nitrogen emission zones in the case of coastal zone nitrogen pollution. The study identified the different sources of nitrogen that could contribute to the low dissolved oxygen content in the Sound and assessed their relative contributions through pathway analysis (Figure 2.2).

Figure 2.2. Sources and pathways of nitrogen in South and Central Puget Sound

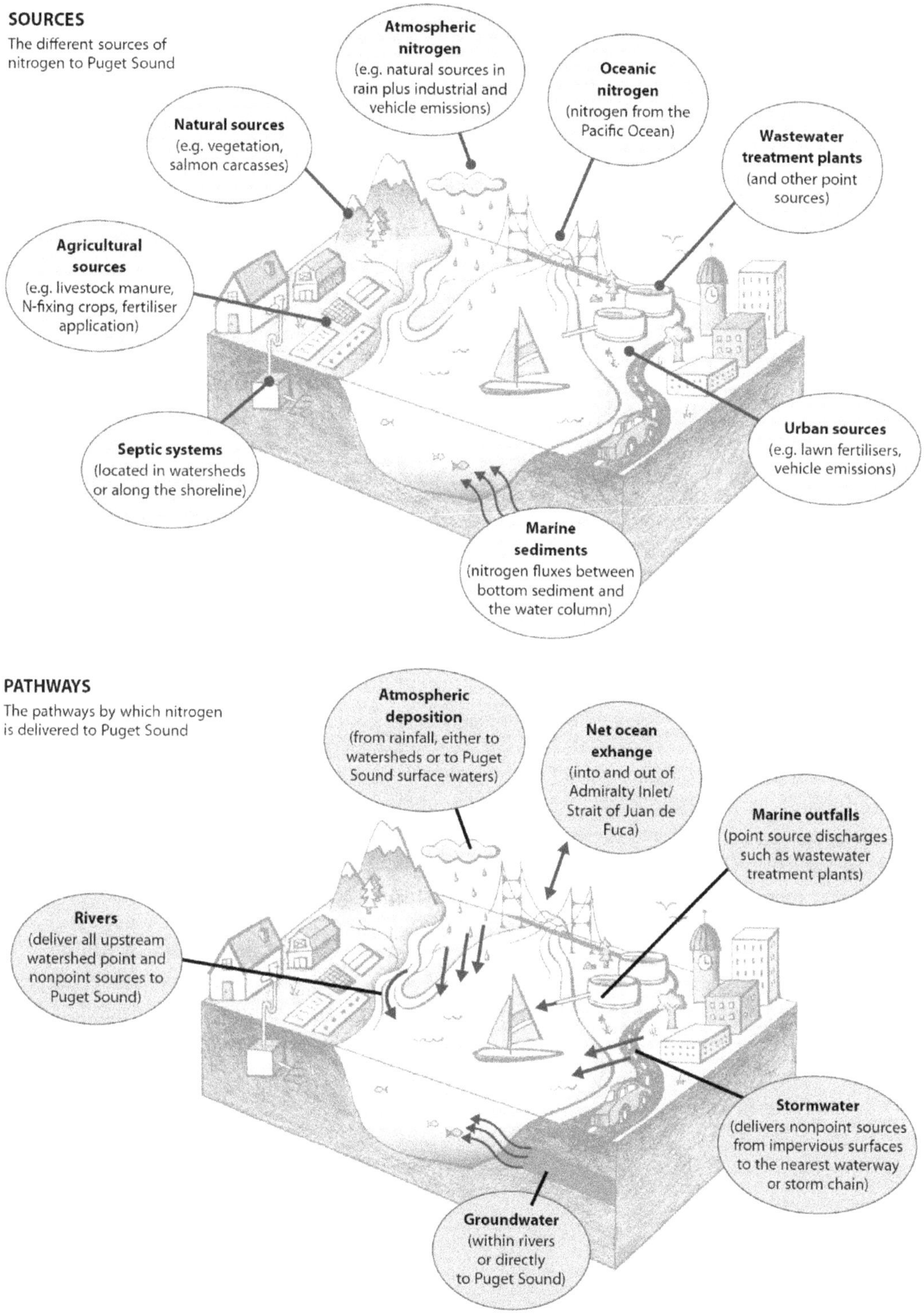

SOURCES
The different sources of nitrogen to Puget Sound

- **Atmospheric nitrogen** (e.g. natural sources in rain plus industrial and vehicle emissions)
- **Oceanic nitrogen** (nitrogen from the Pacific Ocean)
- **Natural sources** (e.g. vegetation, salmon carcasses)
- **Wastewater treatment plants** (and other point sources)
- **Agricultural sources** (e.g. livestock manure, N-fixing crops, fertiliser application)
- **Septic systems** (located in watersheds or along the shoreline)
- **Urban sources** (e.g. lawn fertilisers, vehicle emissions)
- **Marine sediments** (nitrogen fluxes between bottom sediment and the water column)

PATHWAYS
The pathways by which nitrogen is delivered to Puget Sound

- **Atmospheric deposition** (from rainfall, either to watersheds or to Puget Sound surface waters)
- **Net ocean exchange** (into and out of Admiralty Inlet/Strait of Juan de Fuca)
- **Marine outfalls** (point source discharges such as wastewater treatment plants)
- **Rivers** (deliver all upstream watershed point and nonpoint sources to Puget Sound)
- **Stormwater** (delivers nonpoint sources from impervious surfaces to the nearest waterway or storm chain)
- **Groundwater** (within rivers or directly to Puget Sound)

Source: Roberts and Kolosseus (2014).

Another example relates to urban air pollution. In March 2014, Paris suffered a major peak of particle pollution. The scientific assessment revealed that half of the PM_{10} were ammonium nitrate particles formed by a combination of NO_x and NH_3. The NO_x airshed corresponds to the agglomeration (the main source being car traffic). The NH_3 airshed is much larger as emissions come from agricultural activity in north-western France and beyond (see Case Study 1 of Chapter 3.).

The management of air pollution risk in the Santiago Metropolitan Region of Chile is another good example of a policy that takes into account, at least in part, the emission zones of the various pollutants involved.[8] In January 2017, Chile begun implementing a tax on emissions of carbon dioxide (CO_2), PM, NO_x and sulphur dioxide (SO_2). The tax is levied on large stationary sources, particularly fossil fuel-based electricity plants, and takes into account the size of the population affected by the pollution, which is a positive feature. However, the formula considers only the population of the municipality where the source of the emission is located and not, as originally planned, the entire relevant airshed. Since air pollutants can be transported and deposited over relatively large areas, it would have been preferable to take the atmospheric dispersion effects (and the total population exposed) into account when calculating the tax for each polluter (OECD/ECLAC, 2016). Ideally, it would also have been useful to consider other impacts related to deposition, such as exposed ecosystems in the airshed.

IPA is also useful for delineating the risk area. For example, in the German Land of Baden-Württemberg high-resolution deposition analysis and actual monitoring on the ground allowed revealing exceedances of critical loads for eutrophication on major parts of the Land, which previous (lower-resolution and modelling) analysis did not (see Case Study 1 of Chapter 3.).

The second step is to estimate the potential for new or additional emission reductions in the different emission zones, taking into account ongoing emission reduction measures to avoid overlap (Figure 2.3). Cost-effectiveness analysis (CEA) should determine whether to continue (or intensify) existing emission reduction or whether it is preferable to intervene on other emission sources. Denmark is a good example of implementing CEA for managing the risk of eutrophication of coastal waters, lakes and watercourses from agricultural sources of nitrogen. The assessment includes not only an analysis of measures prior to implementation but also a mid-term and a final (ex-post) evaluation (Jacobsen, 2012). For example, the most cost effective measures in the second Danish Action Plan for the Aquatic Environment (1998-2003) were the requirements for catch crops and wetlands, increased utilisation of animal manure and improved feeding practices (Jacobsen, 2004). The least cost-effective measures were land set-aside and increased areas with grass as well as the requirement for reduced animal density (ibid).[9]

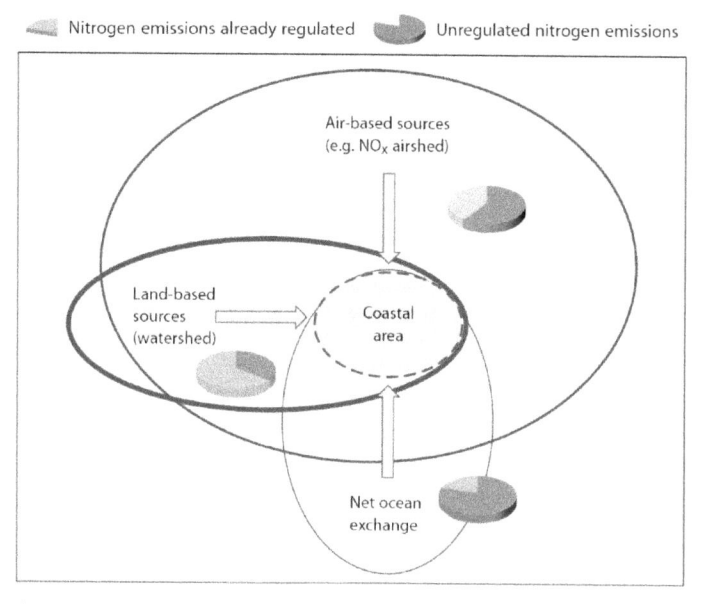

Figure 2.3. State of nitrogen regulation in emission zones

Note: The regulated and unregulated portion of pies is illustrative only.
Source: OECD Secretariat.

The third step in IPA is to compare the cost-effectiveness of emission reductions across all sources of risk (including those already regulated) in all emission zones. This step, which is essential for ensuring coherence of interventions, is still not widespread in the implementation of nitrogen policies. The previous example shows that nitrogen sources other than agriculture – such as industry, wastewater, atmospheric sources and net nitrogen exchanges with marine waters – have not been included in assessing the cost-effectiveness of eutrophication risk management measures in Denmark.

IPA seeks to link the pollution risk to its sources and ensure coherence with ongoing policies to reduce emissions in each source. For example, efforts to reduce NO_x emissions at the national level should be considered in risk areas subject to NO_x deposition. The aim is to select emission sources for which abatement is the most cost-effective, including a further reduction for already regulated nitrogen sources.

To a certain extent, the Baltic Sea Action Plan (BSAP) is a good example of seeking such policy coherence to tackle the risks of eutrophication. Based on scientific advice, the BSAP determines the maximum allowable input (MAI) of nitrogen that the Baltic Sea is thought to be able to tolerate. Each party is then allocated a nitrogen reduction target based on the MAI and according to its watershed area. Finally, atmospheric nitrogen reduction efforts under the LRTAP Gothenburg Protocol are deducted from each party target.

The cost-effectiveness of reducing emissions depends not only on the source of the emission and the form in which the nitrogen is emitted, but also on the place of emission (within the emission zone) and how nitrogen is transported to the risk area. The amount of nitrogen that actually contributes to the risk must be estimated. In fact, only a part of the nitrogen emitted in an emission zone may end up in the risk area.

Typically (according to the US Environmental Protection Agency (USEPA) definition), approximately 75% of the emissions in the airshed of a watershed are redeposited into the watershed, be it NO_x or NH_3. For land-based sources, analysis of nitrogen flows at the watershed level (Billen et al., 2013) suggests that only about 30% of net anthropogenic nitrogen inputs (or NANI)[10] reach the coastal waters. The remaining 70% is either retained by the terrestrial biosphere or denitrified en route, with the relative shares being partly controlled by climatic conditions (Howarth et al., 2012).[11]

The use of models of nitrogen transport in air, land and water can clearly inform cost effectiveness assessments and policy formulation. For example, the U.S. Geological Survey has developed a groundwater vulnerability and assessment (GWAVA) model to relate groundwater NO_3^- concentration observations to spatial attributes representing nitrogen sources and NO_3^- transport and attenuation. GWAVA predicted NO_3^- concentrations in groundwater for the entire United States.[12]

In Denmark, TargetEconN models the attenuation of nitrogen loss from the root zone to the coast with a view to better manage the diffuse pollution of water by nitrogen. This attenuation depends on hydrology, soil types, slope, vegetation and weather conditions. The model was developed to support watershed-scale decision-making. It integrates biophysical and economic modelling. The results clearly indicate that there are large differences in cost-effectiveness between uniform reduction measures and targeted measures, and that more targeted regulation taking account of heterogeneity in both abatement costs and pathways can enhance cost-effectiveness in addressing nitrogen risks (Hasler, 2016).

A third example is NTRADER, which was developed in New Zealand to model the transport of nitrogen from farmland to lakes, including groundwater lag times. NTRADER integrates the outputs of several models, at the farm scale (nitrogen leaching and attenuation model and economic model) and at the catchment scale (nitrogen transport model). It models both nitrogen generation and transport and the economics of "cap and trade" schemes. NTRADER has been useful in managing the risk of water pollution of New Zealand lakes by nitrogen from pastoral lands (Cox et al., 2013).

Case Study 2 of Chapter 3. analyses the management of nitrogen pollution in Lake Rotorua (New Zealand). It also refers to the different models used in the Chesapeake Bay Programme, such as the Community Multi-scale Air Quality model for atmospheric deposition, a watershed transport model, and the Water Quality and Sediment Transport Model for sediment transport.

The ultimate step of the IPA is to estimate the ancillary benefits (avoided damage) of reducing nitrogen emissions in the different emission zones (i.e. taking into account the nitrogen pathways towards the risk area). Indeed, evaluating the benefits of mitigating a tonne of nitrogen released needs to consider the damage avoided in all of the ecosystems through which that tonne would cascade (Moomaw and Birch, 2005). Figure 2.4 provides an example of such "economic cascade" in the Chesapeake Bay watershed.

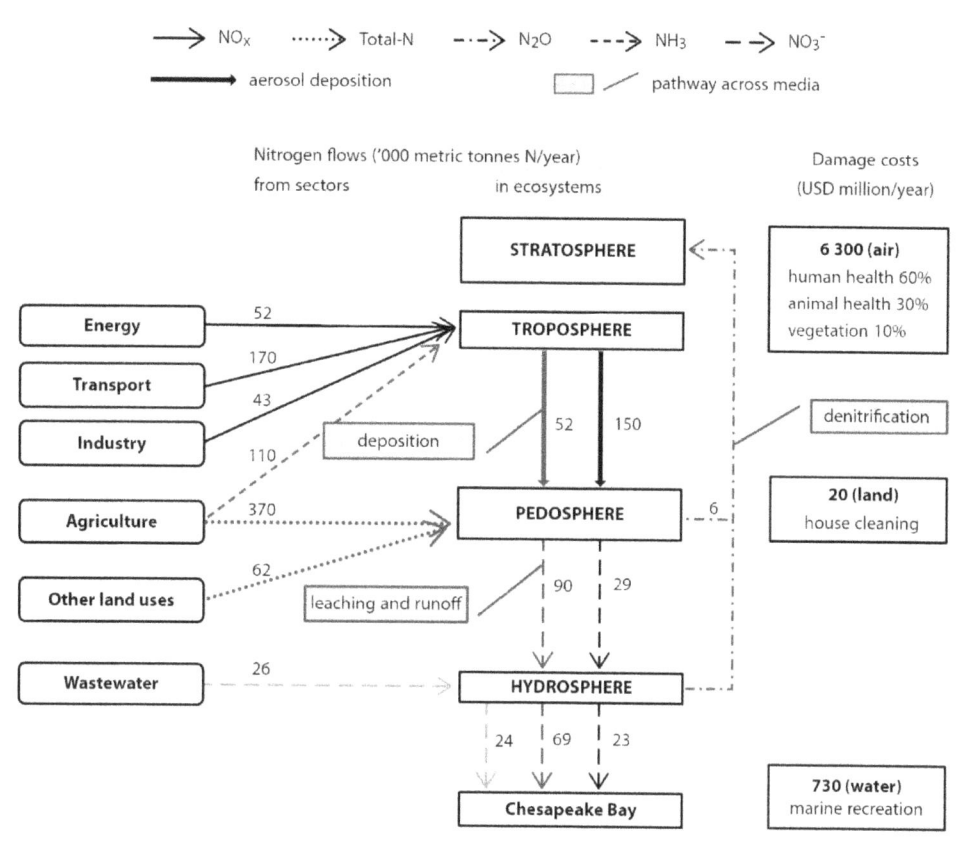

Figure 2.4. Nitrogen flows and cascading damage costs in the Chesapeake Bay watershed

Note: Arrow colour indicates the origin of nitrogen flows: black for air, blue for land and grey for wastewater. Data are illustrative only. Recent estimates indicate that agriculture accounts for 42% of nitrogen mass flows emitted in the Chesapeake Bay watershed, followed by discharges from wastewater (34%) and atmospheric deposition from fossil fuel combustion (24%). These estimates are based on nitrogen monitoring of (i) agriculture, (ii) urban runoff, wastewater and combined sewer overflow, and (iii) atmospheric deposition to forest, non-tidal water and tidal water (www.chesapeakebay.net/indicators/indicator/reducing_nitrogen_pollution, accessed 20 March 2016).
Source: Adapted from Birch et al. (2011).

As can be seen in Figure 2.4, damage costs from air emissions are much larger in the watershed than those from the whole of land and water emissions (despite being smaller nitrogen flows). This reflects human health benefits from reduced air pollution. In addition, since some of the atmospheric nitrogen is found in soils and water, the damage costs are increased accordingly. Thus the reduction of Chesapeake Bay damages from nitrogen (including freshwater and estuarine impacts) may benefit more from a stricter control of air pollution than from stricter water pollution controls. Indeed significant efforts have been made to integrate air and water management in the Chesapeake Bay region (Linker et al., 2013).

Given that the estimated health benefits of cleaner air are much greater than the estimated environmental and health benefits of cleaner water, reducing NO_x and NH_3 emissions will often prove more beneficial than reducing other forms of nitrogen emissions. This is confirmed, for NO_x, by a recent study of nitrogen flows in California's San Joaquin Valley (Figure 2.5). However, this study does not consider the role of NH_3 in the formation of secondary particles ($PM_{2.5}$), which has a significant impact on people's health.

Figure 2.5. Costs of damage and reduction of different forms and sources of nitrogen in the San Joaquin Valley, California

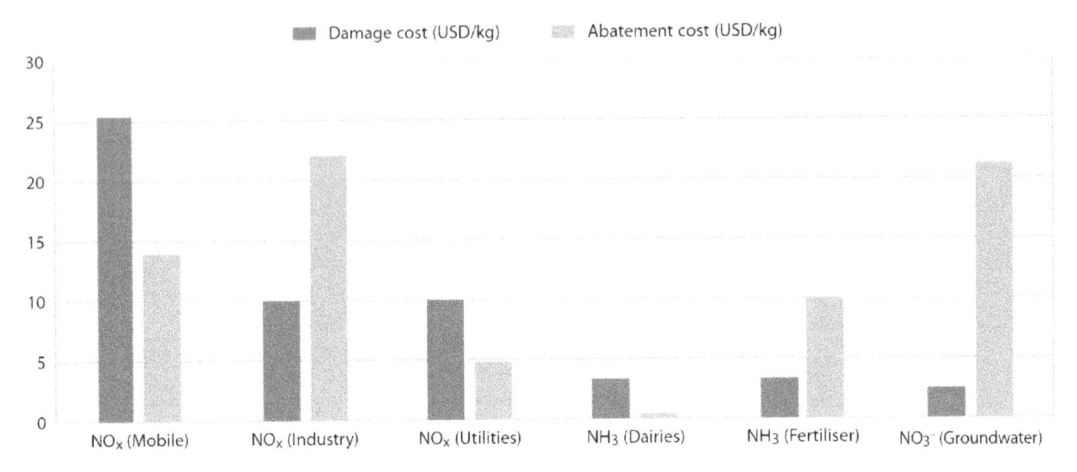

Source: Horowitz et al. (2016).

Assessment of damage avoided in emission zones could consider willingness to pay to improve health and the social cost of carbon. This is what Keeler et al., 2016 did to estimate the social costs of applying a 1kg of fertiliser in Minnesota (Box 2.1). Important assumptions for such valuation are the choices of the value of statistical life (e.g. for premature deaths due to fine particles ($PM_{2.5}$) formation) and the social cost of carbon (for climate related damages). There is also a need to consider the willingness to pay retrospectively (and not only preventively) as it will probably be higher for those who have been drinking contaminated water or breathing polluted air without knowing it.

It can be assumed that each ecosystem responds differently to changes in nitrogen load, depending on the type of ecosystem and local conditions, and that the lag time between emission and impact is specific to each ecosystem. The question arises as to whether a discount factor should be introduced to account for these lag times, as they can significantly influence estimates of damage costs in the IPA (in part, following OECD, 2018). The monetised benefits could be significantly lower, depending on the length of the lag and the discount rate used. Conversely, assuming that the lags do not exist, the value of the benefits of nitrogen reduction will generally be overestimated, perhaps significantly (and critically from the point of view of CBA). For example, Cox et al., 2013 estimate that it takes between 0 and 127 years for nitrogen emitted into the Lake Rotorua watershed to reach the lake via groundwater, depending on whether the source is near or far from the lake (see Case Study 2 of Chapter 3.).

Box 2.1. The social costs of a kilogram of nitrogen fertiliser

A recent study demonstrated that the social cost of a kilogram of nitrogen fertiliser applied in Minnesota varies between less than a tenth of a cent to more than USD 10 depending on the site, the nitrogen form and so-called "end points of interest", that is, whether the impact is about GHG emissions (N_2O), air pollution ($PM_{2.5}$, and indirectly its precursors NO_x and NH_3), or groundwater contamination (NO_3^-) (Keeler et al., 2016).

Following Keeler et al., 2016, it is possible to estimate the social cost of NO_3^- groundwater contamination caused by nitrogen fertiliser. Cost is obtained by multiplying the number of known and predicted contaminated wells by the population using these wells and an average cost of well contamination per household. The latter is estimated on the basis of a survey of well owners facing excessive nitrogen in their wells; it includes the costs of building a new well, buying bottled water or investing in an NO_3^- disposal system. The authors have prepared risk maps for NO_3^- contamination in the various counties of Minnesota by combining, at the county level, data on the three risk factors of agricultural expansion (likelihood), soil characteristics (vulnerability), and population relying on groundwater (exposure).

Keeler et al., 2016, also estimated the social cost associated with NH_3 and NO_x emissions from nitrogen fertiliser, based on their contribution to premature deaths due to $PM_{2.5}$ formation. The cost is obtained by multiplying the number of deaths due to $PM_{2.5}$ downwind of NH_3 and NO_x emissions by an average cost of premature death. An emissions-to-health impact model for $PM_{2.5}$ – the Intervention Model for Air Pollution (InMAP) – calculates the former, while the latter reflects the willingness to pay of people in the United States for a reduction in their mortality risk. InMAP simulates the transport, transformation, and removal of emissions and then calculates mortalities based on resulting $PM_{2.5}$ concentrations, epidemiological information and population census data. NH_3 and NO_x emissions were derived by applying emission factors – 0.08 for NH_3 and 0.005 for NO_x – to the reported on-farm nitrogen inputs in each county. The authors have prepared risk maps for NH_3 and NO_x emissions by modelling damages that occur downwind of their emissions, even beyond the borders of Minnesota, and then allocating the damages back to the county where the NH_3 and NO_x emissions took place.

Finally, Keeler et al., 2016 evaluated the social cost associated with N_2O emissions from nitrogen fertilisers, based on climate related damages. The cost was obtained by converting N_2O into equivalent CO_2 emissions and applying an estimate of the social cost of carbon. This amounts to multiplying the estimated social cost of carbon for CO_2 by 395 (the long-term radiative forcing difference between CO_2 and N_2O). N_2O emissions were derived by applying an emission factor of 0.01 to agricultural nitrogen inputs in each county. For all nitrogen forms (NO_3^-, NH_3, NO_x and N_2O), the social cost per unit of nitrogen was obtained by dividing its total social cost by the agricultural nitrogen inputs in each county.

Beyond IPA, an important question to be clarified for cost-effective management of pollution risk is whether the issue of nitrogen to be addressed is a single-pollutant issue or a multi-pollutant problem. Many of the nitrogen-related impacts fall into the second category and the management of non-nitrogen pollutants may be a priority in some cases. For example, policy to reduce algal blooms may need to address phosphorus (P) first in freshwater lakes, where P limitation is the norm. In these P-limited lakes, it is more effective to initially control the algal growth limiting factor (P intakes) than to seek to reduce excessive nitrogen pollution. The opposite is true in coastal systems, where nitrogen is generally the limiting factor rather than P. The reason is that there is a large amount of denitrification in the coastal zone, so that the equilibrium often passes to an excess of P. Management of both nutrients may sometimes be necessary. This is the case in freshwater lakes, for example, when the reduction of nitrogen inputs improves the composition of algae by reducing harmful cyanobacteria. On the other hand, the reduction of nitrogen inputs in freshwater lakes is ineffective when it leads to an increase in nitrogen-fixing organisms (such as cyanobacteria). Thus, the decision to manage one or both nutrients must be determined on a case-by-case basis.

Focusing on nitrogen pollution only, governments would run the risk of incorrect policy recommendations. IPA is pollutant-specific. It does not seek prioritisation of policy action between nitrogen and other pollutants or precursors. Such prioritisation should ideally entail undertaking pathway assessment(s) for the other pollutant(s) or precursor(s). On the other hand, IPA helps to prioritise between environmental policies, such as between water policy and air policy to manage nitrogen pollution in coastal waters (see Chapter 3. ,Section 3.2.1).

2.1.3 Feasibility of the risk approach

Beyond the criteria of economic efficiency, the "feasibility" of IPA is of course essential for implementation and effective operation of the risk approach (see Chapter 5. for a detailed analysis of the feasibility criteria). This is particularly the case for public acceptability, including the agreement of stakeholders to delineate risk areas and emission zones. For example, in the United States, it would have been appropriate to delineate the entire Willamette River Valley (as a hydrogeological entity) to manage the risk of groundwater NO_3^- contamination in the Willamette Basin, given the high connectivity between the river and shallow groundwater. In practice, only the southern part of the valley could be declared a risk area in the name of the general interest (area used by public systems for the abstraction of drinking water) (see Case Study 2 in Chapter 3.).

Administrative feasibility issues may also arise. Indeed, the nitrogen pathways do not follow the administrative boundaries, far from it. For example, few countries have put in place governance at the level of the airshed (this is more the case for watersheds). The management of emissions in atmospheric basins that have been delineated on a scientific basis can therefore be problematic, as shown by the above example of the Chilean tax on CO_2, PM, NO_x and SO_2 emissions. The Chilean tax takes into account the population of the municipality where the source of the emission is located and not the entire population affected by the pollution (that is, the population of the airshed).

Air and water pollution can extend beyond national borders. In this case, watershed-level or airshed-level emissions management may be problematic in the absence of an international convention or agreement governing transboundary pollution such as LRTAP, for example. IPA has an important role to play in facilitating the adoption of new international provisions to manage transboundary nitrogen-related pollution. The responsibilities and risks of each party could indeed be better identified by the IPA's identification of the emission zones and risk areas.

The feasibility of IPA also raises the issue of its own cost. The preparation of an IPA can involve complex analytical work and require collaboration with a wide range of actors/public authorities, and thus be extremely costly. As a general principle, the level of IPA sophistication should match the expected level of nitrogen pollution risk. When major impacts are at stake, a precise and detailed IPA is required. On the other hand, when risk levels are low, a basic IPA can be used.

2.1.4 The case of nitrous oxide (N_2O)

IPA does not apply to N_2O. As we saw, for a given risk (risk area), the IPA identifies the sources of emission to be managed in priority among a given number of sources, those which are located in the emission zones. In the case of N_2O, the risk area (be it the risk of a greenhouse effect or the risk of depletion of the ozone layer) is global[13] and delineating emission zones, if at all possible, would be of no use for the purposes of CEA (or CBA). Indeed the more sources to compare, the more likely it is to find the one for which emission reduction is the most cost-effective. In other words, as many sources of N_2O as possible should be identified, wherever they are in the country, so as to monitor them and compare the costs – and if possible the ancillary benefits – of reducing their emissions. This requires a global approach.

Such global approach aims to manage only the sources of N_2O unlike a precautionary approach, as described in Section 2.2, which covers all nitrogen forms. As we have seen (Chapter 1.), agriculture accounts for the majority (about 2/3) of anthropogenic emissions of N_2O and N_2O emission data by source are available (see FAOSTAT for example). However, because of the complexities of farm biological and management systems, accuracy of N_2O emission data is limited (PCE, 2016). For example, current research in New Zealand is looking at improving a modelling tool[14] to enable more accurate measurement of emissions from individual farms and thus better estimation of the impacts of different farm management practices on N_2O emissions (ibid).

It is therefore necessary to better understand the nitrogen pathways in the soil to better identify the sources of N_2O and the level of their emissions.[15] This involves estimating the share of incomplete denitrification (that is to say the ratios N_2O/dinitrogen) (see Section 1.4.5). The difficulty of obtaining accurate N_2O emission data arises in part from the complexity of the biological systems involved. Denitrification involves more than 150 known species of bacteria (see Annex A). Although it is well established that soils are a dominating source for N_2O, researchers are still struggling to fully understand the complexity of the underlying microbial production and consumption processes (Butterbach-Bahl et al., 2013).[16] The processes of N_2O formation in the oceans are no less complex. For example, recent research revealed that large amounts of N_2O are produced in regions of the Atlantic Ocean with little oxygen[17] (Grundle et al., 2017).

Applying a global approach does not necessarily imply that governments should seek a goal of reducing N_2O emissions. The climate change mitigation policy does not set an individual reduction target for each GHG. Instead, N_2O is part of a GHG basket under the UN Framework Convention on Climate Change (UNFCCC) and countries can decide how to prioritise GHG emission reductions in their own nationally determined contributions. Achieving the temperature goals of the Paris Agreement will require net zero (or lower) emissions of long-lived GHGs in the second half of the century (see for example, Rogelj et al., 2015). The implications for mitigation of N_2O are as yet unclear and, as with other GHGs, also depend on the potential for natural and artificial carbon sequestration. Meanwhile, N_2O concentrations in the atmosphere continue to increase (Figure 2.6).

Figure 2.6. Global average atmospheric concentrations of nitrous oxide (N_2O)

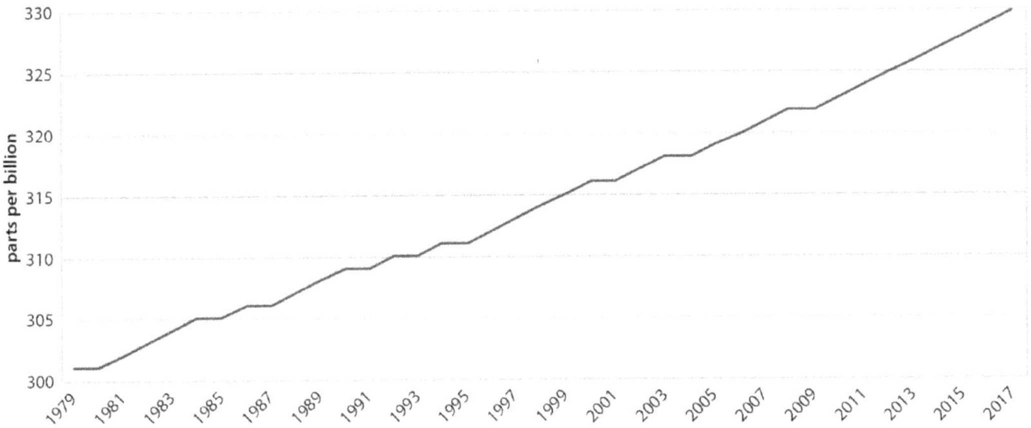

Source: The National Oceanic and Atmospheric Administration (NOAA) Annual Greenhouse Gas Index (AGGI), Spring 2018, https://www.esrl.noaa.gov/gmd/aggi/aggi.html, accessed 9 July 2018.

2.2 The "precautionary" approach

The increasing amount of nitrogen on Earth has enhanced the speed of nitrogen cycling in the environment, i.e. both the rate at which nitrogen has been added to but also lost from the environment has increased (Müller and Clough, 2014). The uncertainties associated with such an acceleration of the nitrogen cycle and their implications for the nitrogen cascade pathways raise the question of the precautionary principle (PP).

There is no definitive statement of the PP, but there is a reasonable consensus about what it says, at least among its proponents (in part, following Saunders, 2010). When an activity raises threats of harm to human health or the environment, precautionary measures should be taken even if some cause and effect relationships are not fully established scientifically. In other words, the principle is to be applied when (a) there is scientific evidence for a threat to the environment or to health, but (b) the evidence, while sound, is not conclusive. According to the European Commission, the PP applies "where preliminary objective scientific evaluation indicates that there are reasonable grounds for concern" (EC, 2000). Under the PP, it is (somehow) deemed that there is not enough science to conduct an IPA. What is crucial, however, is that there must be a prima facie scientific case for a threat. If this is not the case, the application of the principle is not justified in any way.[18]

According to Battye et al., 2017, the rapid increase in human production of nitrogen – by almost five-fold in the last half-century – "has removed any uncertainty about the importance of human-produced nitrogen on the overall nitrogen cycle". Battye et al., 2017, questions whether denitrification can continue to keep pace with the increase in human production of nitrogen, which seems to be the case so far. For example, if the habitats of denitrifying bacteria, such as marshes and wetlands, were to be reduced, a major imbalance in the nitrogen cycle could occur. Alteration of the denitrification process could lead to a chain reaction on nitrogen pathways with unpredictable but potentially disastrous consequences on health and the environment. If not, if denitrification can meet the growing demand, then the environmental consequences are predictable and are a matter of risk management (for example, risk of increasing the greenhouse effect and depleting the ozone layer).

A second key criterion for triggering the implementation of the PP is that the proponent of an activity, rather than the public, must bear the burden of proof. It is for the scientific community to demonstrate prima facie that the acceleration of the nitrogen cycle will result in negative effects that are not already taken into account by environmental policy and that cannot be because they are not linked to one but to all environmental media. In other words, it is necessary to demonstrate that there is a systemic effect linked to the nitrogen cascade.

A third factor to consider is that the PP, and therefore the management of uncertainty, must be viewed within the overall framework of risk management (Box 2.2) According to the European Commission, the PP "forms part of a structured approach to the analysis of risk, as well as being relevant to risk management" (EC, 2000). Thus, a poorly defined PP may "direct resources toward attempts to control poorly understood, low-level risks using resources that could be more effectively directed toward the reduction of well-known, large-scale risks" (Majone, 2010).

Box 2.2. Differentiating between risk and uncertainty

A common distinction between risk and uncertainty derives from Knight's (1921) observation that risk is uncertainty that can be reliably measured. Thus, risk describes the likelihood and consequence of an uncertain event of which the probability of occurrence can be reliably estimated. Uncertainty describes situations where the probability of occurrence is not known and perhaps cannot be known. The difference between risk and uncertainty can be understood as a spectrum, where uncertainty is an expression of the degree to which a value or relationship is unknown.

Pollutants have been defined by Holdgate, 1979, as substances causing damage to receptors in the environment (in part, following IEEP, 2014). The pollutant may be emitted from a "source" into the environment, through which it travels along a "pathway" till it reaches a receptor and creates an "impact". It follows from this definition that if the pollutant reaches no receptor in damaging quantities – because it has been rendered harmless along the cascade either by being transformed into another substance or into a form where it cannot affect the receptor or because it has been diluted to harmless levels – then there has been no pollution. It also follows that as the mere emission of a potential pollutant to the environment does not necessarily constitute pollution, managing uncertainty does not necessarily require a reduction in all emission sources. It requires consistency with risk management. The PP should therefore be closely associated with the risk approach as part of a dual approach to managing human impacts on the nitrogen cycle. It must complement the risk approach and not replace it. It must aim to limit the total nitrogen load entering the system and, where appropriate, propose measures in addition to and in line with risk management measures.

This leads to the question of cost-effectiveness, the fourth criterion to be taken into account in applying the PP. This criterion was already embedded in Principle 15 of the Rio Earth Summit Declaration on Environment and Development, which emphasised "where there are threats of serious or irreversible damage, lack of full scientific certainty shall not be used as a reason for postponing cost-effective measures to prevent environmental degradation" (UNGA, 1992). Managing uncertainties of the nitrogen cascade for a given country means covering the entire economy. In comparison to environmental policies that target sources according to impacts, the precautionary approach offers flexibility in the choice of nitrogen sources to manage (for example, among agriculture, transport, energy, industry, wastewater). While such flexibility is synonymous with potential cost-efficiency gains, this is not the case in terms of environmental effectiveness.

The scientific community is calling for reductions in nitrogen emissions and improved efficiency of nitrogen use (see for example the concept of planetary boundaries discussed below). The USEPA's Science Advisory Board estimated that approximately 7 million tonnes of nitrogen could be reduced each year (nearly 25% of current nitrogen emissions in the United States) by disseminating technologies available in precision agriculture, fertilisers, NO_x control, wetland creation and sewage treatment (USEPA-SAB, 2011). Jörß et al., 2014 and Döhler et al., 2011 see significant reduction potentials in the German energy and agriculture sectors; SRU, 2015 estimated the overall nitrogen reduction potential in the German agricultural sector at around 40% of current emissions. Sutton et al., 2013 and Tomich et al., 2016 suggest improving the efficiency of nitrogen use along food and energy supply chains. Seitzinger and Phillips, 2017 wonder whether nitrogen use efficiency could not constitute a kind of "heuristic device", a proxy to guide policy making. However, pointing out potentials of reduction or gain in efficiency in the use of nitrogen is not sufficient. Policies must be designed to achieve these reductions or efficiencies in the use of nitrogen in a cost-effective manner, i.e. not just in terms of cost but also in relation to environmental impacts.

For example, reducing nitrogen emissions in the San Joaquin river watershed can be cost-efficient in reducing the total amount of nitrogen emitted in the state of California, given the importance of this "specific high-nitrogen region" for nitrogen emissions in California.[19] But this is certainly not the most effective way to improve the quality of groundwater in the Central Valley aquifer system and of coastal waters in the San Francisco bay (Figure 2.7) not to mention air quality in the City of Los Angeles. These risks must be managed at different geographical scales by specific risk measures and not by untargeted actions whose sole purpose is to reduce the overall nitrogen load in California. In other words, risk management and precautionary measures are not mutually exclusive, they must be complementary.

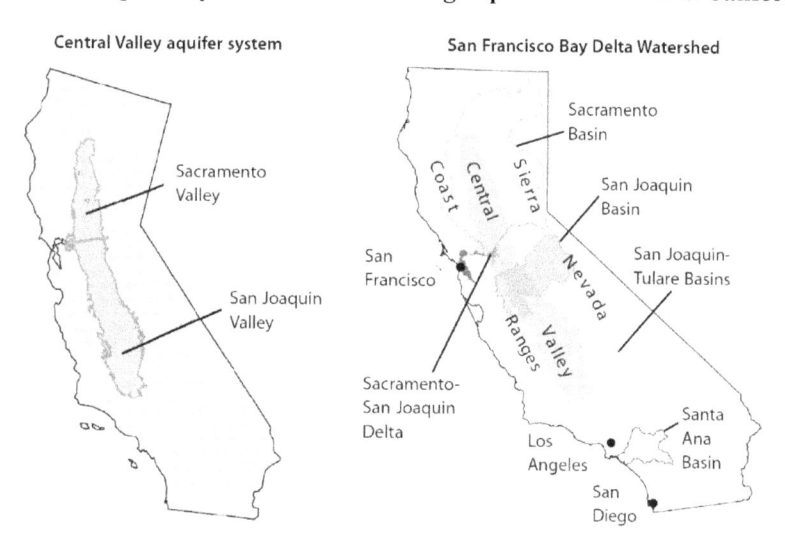

Figure 2.7. Aquatic systems at risk of nitrogen pollution in central California

Left map. Intense demand for water in the Central Valley of California and related increases in groundwater nitrate (NO$_3^-$) concentration threaten the sustainability of the groundwater resource (Ransom et al., 2017). The Central Valley of California, including the Sacramento Valley and the San Joaquin Valley, must be considered when assessing groundwater contamination risk in the region.
Right map: The San Francisco Bay Delta Watershed consists of several major waterways, including the Sacramento and San Joaquin Rivers and their tributaries. Where these two large rivers meet near Sacramento, a great inland Delta is formed where the river waters collect before passing into San Francisco Bay. The Sacramento and San Joaquin River Basins and the Sacramento-San Joaquin Delta should be considered when assessing the risk of nitrogen pollution of the coastal areas into which the San Francisco Bay Delta Watershed is draining, including the San Francisco Bay, San Pablo Bay, Suisun Bay, and the Golden Gate Strait.
Source: Ransom et al. (2017) (Left map) ; Kratzer et al. (2011) (Right map).

Fifth, the economic and policy implications of using the PP need to be carefully analysed. Use of the principle with respect to the nitrogen cycle can influence production decisions (for example, farmers switching to crops with low nitrogen inputs)[20] and trade (e.g. through restricting imports of foods produced with high inputs of nitrogen). Applying the PP to the nitrogen cycle can foster the development of new technologies, the dissemination of which is itself put into question by the principle, as is the case of biotechnology (genetically modified organisms). For example, it could lead to an increase in genetic engineering research to develop new nitrogen-fixing crops[21] to increase crop productivity while reducing fertiliser use.[22]

Precautionary management to cope with the uncertainties of the nitrogen cascade finally raises the thorny question of the limit to be set, the level acceptable to society, in terms of the net nitrogen balance of a country, or even of the planet. To help decision-making, some studies have estimated "boundaries" or "tipping points" for each form of nitrogen beyond which "critical limits" of their respective impacts would be exceeded (Box 2.3). As these studies acknowledge, such estimates should not lead to setting global emission reduction limits, given the spatial (and temporal) variability of nitrogen impacts on air, water, biodiversity and soils and, for N_2O, the possibility of mitigating global warming by acting on other GHGs. The boundary concept is even less conducive to setting a single threshold for nitrogen in all its forms. Indeed, there is no such thing as a single value of the damages caused by nitrogen – or a fixed exchange rate between the emissions of different forms of nitrogen – since the impacts of nitrogen are multiple and are specific to the site and form of nitrogen.[23]

Box 2.3. Estimating boundaries for different nitrogen forms

The concept of planetary boundaries

The concept of planet-wide environmental boundaries, or "tipping points", has recently been introduced in an attempt to respond to the multiple pressures on the Earth's system caused by the acceleration of human activities (Rockström et al., 2009).[24] The Planetary Boundaries (PB) framework identifies acceptable levels of anthropogenic perturbations below which the risk of destabilisation of the earth system is likely to remain low – a "safe operating space" for global societal development. Beyond the boundaries, the earth system is at risk, and with it humanity. In this way, PBs provide a science-based analysis of the risk to destabilise the earth system at the planetary scale.

Rockström et al., 2009 proposed nine PBs, namely climate change, biodiversity loss, increased nitrogen and phosphorus cycling, stratospheric ozone depletion, ocean acidification, global freshwater use, change in land use, chemical pollution and atmospheric aerosol loading. Three PBs were identified as having already exceeded their tipping points – climate change, biodiversity loss and human interference with the nitrogen cycle. On the latter, the safe operating space for anthropogenic nitrogen fixation was roughly estimated at 35 million tonnes N per year, or 17% of its current value as estimated by Fowler et al., 2013 (see Table 1.1 in Chapter 1). Rockström et al., 2009 admitted the roughness of this estimate, implying the need for an update.

First criticism

The concept of PB appeared to be controversial as it was deemed too low to feed the current world population (Nordhaus et al., 2012). Since PBs aim to ensure a "safe operating space" for human development, the human need for nitrogen should be considered as well the environmental impacts.

In response to this first criticism, de Vries et al., 2013 estimated global food production needs at ~ 52-80 million tonnes N per year and losses to the environment deemed 'acceptable' (i.e. below critical thresholds for health and the environment) at 20-133 million tonnes N per year[25] (Table 2.2). Although it is two to six times higher than the previous estimate (72 to 213 million tonnes N per year compared to 35 million N tonnes per year), this new estimate of PB remains below or close to the current rate of anthropogenic fixation estimated by Fowler et al., 2013 at 210 million tonnes N annually. In other words, according to de Vries et al., 2013 even if an optimal allocation of nitrogen (and phosphorus) could be achieved across the planet, it is likely that the PB for nitrogen is lower than the current fixation.

Table 2.2. Planetary boundaries for anthropogenic nitrogen fixation

Million tonnes of N per year

Indicator	Critical limits	Current nitrogen losses	Planetary boundaries
NH_3 concentration in air[1]	1 µg per m³	24.9	89
	3 µg per m³	32.1	115
Radiative forcing of N_2O[2]	1 W per m²	0.8	20
	2.6 W per m²	5.3	133
NO_3^- concentration in drinking water[3]	25 mg per litre	30.0	83
	50 mg per litre	36.9	111
Nitrogen in surface water[4]	1.0 mg N per litre	5.4	62
	2.5 mg N per litre	7.2	82
Phosphorus in water[5]	6.2 million tonnes P per year	-	73
	11.2 million tonnes P per year	-	132
Global needs of nitrogen for food production[6]	Current NUE	-	80
	25% increase in NUE	-	52

1. Upper and lower limits for impacts on lichens and higher plants, respectively
2. Using the present share of N_2O to the sum of CO_2, methane (CH_4) and N_2O and excluding the effects of nitrogen on CO_2 sequestration
3. World Health Organisation (WHO) drinking water limits
4. Critical limits in terms of surface water eutrophication (dissolved inorganic nitrogen)
5. Critical limits in terms of phosphorus fertiliser input (Carpenter and Bennett, 2011)
6. For 9 billion people consuming food at the recommended level, thus avoiding both overconsumption and malnutrition, assuming either current Nitrogen Use Efficiency (NUE) in agriculture or a 25% increase in NUE

Source: After De Vries et al. (2013); Steffen et al. (2015).

Second criticism

The second, more fundamental, criticism of the PB concept is the irrelevance of a global threshold due to spatial variability (Lewis, 2012; Nordhaus et al., 2012). First, many impacts occur at a regional level (e.g. terrestrial biodiversity decline due to nitrogen deposition; eutrophication of fresh and marine waters due to nitrogen runoff). Second, the supply of nitrogen (and phosphorus) fertilisers is very unevenly distributed between OECD countries and the rest of the world (Vitousek et al., 2009).

Some scientists have argued that this spatial variability of nitrogen impacts and nitrogen availability would be taken into account by establishing regional boundaries (RBs), citing as an example of RB the critical loads of nitrogen deposition. According to Vries et al., 2001 and Erisman et al., 2001, the setting of RBs does not preclude assessing boundaries on a larger scale, as was done in the Netherlands; exceeding these "upper boundaries" would signal potential regional problems unrelated to the spatial allocation of nitrogen.

Overall assessment

A PB for N_2O has little policy relevance; instead, any radiative forcing boundary must refer to all GHGs as the reduction of N_2O emissions can be exchanged for CO_2 or CH_4 reductions. RBs are already part of nitrogen risk management (such as the risks of acidification and eutrophication in the example of critical loads of

> deposition on terrestrial ecosystems). Pending more work on the relevance of establishing PBs or RBs, as in the framework of the International Nitrogen Management System (INMS), a national nitrogen budget could be a starting point for any precautionary approach (see below).

It is too early to discuss any limit or level of efficiency improvement desirable in the use of nitrogen or its losses. The ongoing work of the INMS is expected to shed light on this issue by 2021 (Box 2.4). In the meantime, as a first step, countries could establish an economy-wide nitrogen balance and begin to monitor trends. This would involve assessing the total amount of nitrogen introduced into the environment from all sources and monitoring these sources in order to report – both by source and overall – the amount of nitrogen released each year, also taking into account denitrification. This could be undertaken in parallel with risk-based efforts to manage specific nitrogen impacts.

Box 2.4. The International Nitrogen Management System

The Global Environment Fund (GEF) has pledged close to USD 6 million to set up the INMS (via the so-called "Toward INMS" process). The four-year timeframe for Toward INMS is 2017-21. Coverage is global. UNEP is the GEF 'Implementing Agency' (i.e. policy customer), while the UK Natural Environment Research Council, Centre for Ecology and Hydrology (CEH) is the GEF 'Executing Agency' (i.e. project coordinator). UNEP contribution to Toward INMS is through providing the Secretariat function of the Global Partnership on Nutrient Management (GPNM), which was established in 2007 to steer dialogue and actions to promote effective nutrient management.

INMS will provide a combination of analyses to support nitrogen policy making. The Toward INMS architecture includes four components: (i) tools for understanding and managing the global nitrogen cycle; (ii) quantification of nitrogen flows, threats, benefits; (iii) regional demonstrations; and (iv) awareness raising and knowledge sharing.

Seven regional demonstrations have been set up to show the benefits of joining up nitrogen management at the regional/catchment scale:

- Western Europe (Atlantic coast, including France, Spain and Portugal)
- East Asia (including China and Japan)
- South Asia (including India, Maldives and Sri Lanka)
- Eastern Europe (Dnieper catchment, including Russia and Ukraine)
- Latin America (La Plata catchment, including Brasil)
- East Africa (Lake Victoria catchment, including Uganda)

Monitoring of nitrogen mass flows (e.g. via an economy-wide national nitrogen balance) would provide an early-warning indicator of whether policies are easing the overall situation or are only shifting the problem from one location to another

(Salomon et al., 2016). In other words, if the nitrogen released into the environment has not been significantly reduced by the policies in place, this may indicate that not all impacts have been satisfactorily managed (i.e. that nitrogen policies have not all been effective) or that new risks are incurred. Such "headline indicator" would also raise policy makers' awareness to the systemic dimension of the nitrogen issue and would improve policy coherence (ibid).

Some countries have undertaken such comprehensive monitoring of their nitrogen mass flows, such as Denmark, Germany, The Netherlands, Switzerland and the United States. The European Nitrogen Assessment (Sutton et al., 2011) provides estimates for the EU-27 and the California Nitrogen Assessment estimates a state-level economy-wide nitrogen balance (Tomich et al., 2016). In 2012, the OECD Working Party on Environmental Information (WPEI) initiated discussions on the methodology for measuring nitrogen flows with a view to developing a country-wide nitrogen indicator. The WPEI proposed a tiered measurement framework and a reporting template that could form the basis for defining nitrogen indicators for inclusion in the OECD sets of green growth and environmental indicators (OECD, 2013 and 2014).

Notes

[1] What Keeler et al., 2016 call the "end points of interest".

[2] This report does not detail the acceptable levels of risks set by environmental policy.

[3] In December 2015, the twenty-first session of the Conference of the Parties (COP 21) to the UN Framework Convention on Climate Change (UNFCCC), held in Paris, set a long-term goal of keeping the increase in global average temperature to well below 2°C above pre-industrial levels.

[4] As estimated by using carbon (C) stock information in the national inventory of GHG emissions and sinks. Nitrogen stock change was determined by simply assigning a molar C/N ratio of 12 for soils and 261 for trees and making the appropriate conversions from C to N.

[5] The literature review of Battye et al., 2017 leads to the same conclusion: changes in terrestrial biomass and marine sediments, which are the main nitrogen sink after denitrification, would capture only 9 and 13 Tg N per year respectively, compared to 400 Tg N per year for denitrification.

[6] Biogeochemical pathways are referred to in this report as "pathways".

[7] As defined by the USEPA, an airshed defines the geographic area that contains the emissions sources that contribute 75% of the nitrogen form (e.g. NO_x) deposited in a particular watershed, city or protected ecosystem.

[8] Air pollution is considered the biggest environmental challenge in Chile, particularly fine particles ($PM_{2.5}$) pollution in the Metropolitan Region of Santiago, two thirds of which are secondary particles with NO_x as one of the precursors.

[9] Other benefits are not included in the calculation, and this is the main reason why area-related measures generally have the lowest cost efficiency.

[10] The net input term of NANI includes deposition of NO_y (the sum of all oxidised nitrogen forms), nitrogen fertiliser use, nitrogen fixation by crops and net nitrogen in import and export of food and feed. The output term is riverine export (Hong et al., 2013).

[11] The share remitted back to the atmosphere is significantly higher than the share absorbed by terrestrial plants, according to USEPA-SAB, 2011.

[12] GWAVA consists of two nonlinear regression models: one for shallow groundwater (typically < 5 m deep, and which may or may not be used for drinking) and the other for deeper supplies used for drinking.

[13] No matter where N_2O is emitted, it will contribute to climate change and ozone layer depletion effects globally (see Chapter 1).

[14] The modelling tool, named Overseer, was originally designed for managing nitrogen fertiliser; it can also be used to estimate N_2O emissions by combining GHG emission factors used in the national inventory with farm data.

[15] A better understanding of the pathways would also be required for estimating the share of N_2O emissions that contribute to ozone layer depletion.

[16] For example, current biochemical models hold that inorganic hydroxylamine is the only intermediary formed when nitrifying bacteria convert ammonium (NH_4^+) into dormant nitrite (NO_2^-). In a new study, chemists found that hydroxylamine is converted into another intermediary – nitric oxide (NO) - which under normal soil conditions acts as the

chemical prelude to NO_2^-; but under imperfect soil conditions, NO is converted into N_2O (Caranto and Lancaster, 2017).

[17] Extreme low oxygen concentrations occur in the Atlantic in ocean eddies of up to 100 kilometres in diameter.

[18] So far, there has been a narrow and selective application of the PP. There are only two cases where it has been invoked: climate change and biotechnology products, including the use of genetically modified organisms (GMOs) to reduce the use of herbicides (Saunders, 2010).

[19] The California Nitrogen Assessment has chosen to study the San Joaquin river watershed "because of the mix of intensive agricultural production combined with large urban areas" and a "history of air and water pollution associated with excess nitrogen" - a so-called "specific high-nitrogen region" (Tomich et al., 2016). Groundwater pollution is described as the main nitrogen pollution problem.

[20] For example, grasses such as chicory, plantain, red clover or white clover may be substituted for perennial ryegrass as forage in temperate regions, with perennial ryegrass having a very high need for nitrogen to support its growth (Gilliland et al., 2010).

[21] Either through a transgenic strategy (i.e. by transferring the nitrogenase gene from the bacteria to the plant) or by extending the symbiotic fixation of nitrogen to non-legumes (i.e. by letting the bacteria fix the nitrogen, but by making it happen inside the plant instead of the soil).

[22] Nitrogen fixed by legumes responds more to the needs of the host plant than spreading chemical fertilisers, no matter how precise.

[23] Reasoning by analogy with the possibility of setting a carbon price based on the global warming potential of different GHGs, Socolow, 1999 and 2016 proposes to use a single price of nitrogen to promote an efficient use of nitrogen.

[24] The acceleration of human activities has been so dramatic that a new geological epoch, the Anthropocene, has been proposed (Crutzen, 2002; Crutzen and Steffen, 2003).

[25] Steffen et al., 2015 estimated a PB for phosphorus (P) based on the need of P in plant growth, using an average N:P ratio in growing plant tissue of 11.8.

References

Battye, W. et al. (2017), "Is Nitrogen the Next Carbon?", *Earth's Future*, 5(9).

Billen, G. et al. (2013), "The Nitrogen Cascade from Agricultural Soils to the Sea: Modelling Nitrogen Transfers at Regional Watershed and Global Scales", *Phil. Trans. R. Soc. B*, 368(1621), doi.org/10.1098/rstb.2013.0123.

Birch, M.B. et al. (2011), "Why Metrics Matter: Evaluating Policy Choices for Reactive Nitrogen in the Chesapeake Bay Watershed", *Environmental Science & Technology.*, Vol. 45, N°1.

Butterbach-Bahl, K. et al. (2013), "Nitrous Oxide Emissions from Soils: How Well do we Understand the Processes and their Controls?", *Phil Trans R Soc B*, 368: 20130122, doi.org/10.1098/rstb.2013.0122.

Caranto, J. D. and K.M. Lancaster (2017), "Nitric Oxide is an Obligate Bacterial Nitrification Intermediate Produced by Hydroxylamine Oxidoreductase", *Proc Natl Acad Sci USA*, 114(31), doi.org/10.1073/pnas.1704504114.

Cox, T.J. et al. (2013), "An Integrated Model for Simulating Nitrogen Trading in an Agricultural Catchment with Complex Hydrogeology", *Journal of Environmental Management*, 127 (2013), doi.org/10.1016/j.jenvman.2013.05.022.

Crutzen, P.J. (2002), "Geology of Mankind", *Nature*, 415.

Crutzen, P.J and W. Steffen (2003), "How Long Have We Been in the Anthropocene Era?", *Climatic Change*, 61 (3).

de Vries, W. et al. (2001), "Assessment of Nitrogen Ceilings for Dutch Agricultural Soils to Avoid Adverse Environmental Impacts", *The Scientific World*, 1 (S2).

de Vries, W. et al. (2013), "Assessing Planetary and Regional Nitrogen Boundaries Related to Food Security and Adverse Environmental Impacts", *Current Opinion in Environmental Sustainability*, 5 (3–4).

Döhler, H. et al. (2011), "Systematische Kosten-Nutzen-Analyse von Minderungsmaßnahmen für Ammoniakemissionen in der Landwirtschaft für Nationale Kostenabschätzungen", Environmental Protection Agency, Dessau-Roßlau UBA-Texte 79/2011.

EC (2000), "Commission adopts Communication on Precautionary Principle", European Commission, IP/00/96, europa.eu/rapid/press-release_IP-00-96_en.htm.

Erisman, J.W. et al. (2001), "An Outlook for a National Integrated Nitrogen Policy", *Environmental Science and Policy*, 4.

Fowler, D. et al. (2013), "The Global Nitrogen Cycle in the Twenty-first Century", *Phil. Trans. R. Soc. B*, 368(1621).

Gilliland, T.J. et al., (2010), "The Effect of Grass Species on Nitrogen Response in Grass Clover Swards", *Advances in Animal Biosciences*, 1(1), Proceedings of the British Society of Animal Science and the Agricultural Research Forum.

Grundle, D.S. et al. (2017), "Low Oxygen Eddies in the Eastern Tropical North Atlantic: Implications for N_2O Cycling", *Scientific Reports*, 7(4806), doi.org/10.1038/s41598-017-04745-y.

Hasler, B. (2016), "Spatial Modelling of Non-Point Nitrogen Loads to Fulfil Water Quality Policies", paper presented at the European Association of Environmental and Resource

Economists, 22nd Annual Conference, 22 - 25 June 2016, Zurich, Switzerland, www.webmeets.com/files/papers/EAERE/2016/1208/Spatialmodellingnitrogenabatement.pdf.

Holdgate, M W (1979), *A Perspective of Environmental Pollution*, Cambridge University Press.

Hong, B. et al. (2013), "Estimating Net Anthropogenic Nitrogen Inputs to U.S. Watersheds: Comparison of Methodologies", *Environ. Sci. Technol*, 47.

Horowitz, A.I. et al. (2016), "A Multiple Metrics Approach to Prioritizing Strategies for Measuring and Managing Reactive Nitrogen in the San Joaquin Valley of California", *Environmental Research Letters*, 11 (6), doi.org/10.1088/1748-9326/11/6/064011.

Howarth, R. et al. (2012), "Nitrogen Fluxes from the Landscape are Controlled by Net Anthropogenic Nitrogen Inputs and by Climate", *Frontiers in Ecology and the Environment*, *10*(1), doi.org/10.1890/100178.

IEEP (2014), *Manual of European Environmental Policy*, Institute for European Environmental Policy, Farmer, A.M. editor, Routledge, London.

Jacobsen, B.H. (2012), "Analyse af landbrugets omkostninger ved implementering af vandplanerne fra 2011", [Economic Evaluation of the Agricultural Costs Related to the River Basin Management Plans from 2011], FOI Udredning; Nr. 2012/6, Danish Research Institute of Food Economics, University of Copenhagen.

Jacobsen, B.H. (2004), *Vandmiljøplan II – økonomisk slutevaluering*, [Economic Evaluation of Action Plan on the Aquatic Environment II], Report N° 169, Danish Research Institute of Food Economics, University of Copenhagen.

Jörß, W. et al. (2014), Luftqualität 2020/2030: Weiterentwicklung von Prognosen für Luftschadstoffe unter Berücksichtigung von Klimastrategien. Environmental Protection Agency, Dessau-Roßlau UBA-Texte 35/2014.

Keeler, B.L. et al. (2016), "The Social Costs of Nitrogen", *Science Advances*, 2(10).

Knight, F.H. (1921), *Risk, Uncertainty, and Profit*, Hart, Schaffner and Marx, Houghton Mifflin Company, Boston.

Kratzer, Ch. R. et al. (2011), *Trends in Nutrient Concentrations, Loads, and Yields in Streams in the Sacramento, San Joaquin, and Santa Ana Basins, California, 1975–2004*, Scientific Investigations Report 2010–5228, National Water-Quality Assessment Program, U.S. Geological Survey.

Lewis, S.L. (2012), "We Must Set Planetary Boundaries Wisely", *Nature*, 485.

Linker, L.C. et al. (2013), "Computing Atmospheric Nutrient Loads to the Chesapeake Bay Watershed and Tidal Waters", *Journal of the American Water Resources Association*, DOI: 10.1111/jawr.12112.

Majone, G. (2010), "Strategic Issues in Risk Regulation and Risk Management", in *Risk and Regulatory Policy: Improving the Governance of Risk*, OECD Publishing, Paris, doi.org/10.1787/9789264082939-7-en.

Moomaw, W.R. and M.B. Birch (2005), "Cascading Costs: an Economic Nitrogen Cycle", *Science in China Series C: Life Sciences*, 48(2) Supplement.

Müller, C and T. J. Clough (2014), "Advances in Understanding Nitrogen Flows and Transformations: Gaps and Research Pathways", *Journal of Agricultural Science*, Special Issue from the 17th International Nitrogen Workshop, 152 (S1).

Nordhaus, T. et al. (2012), *The Planetary Boundaries Hypothesis a Review of the Evidence*, The Breakthrough Institute, Oakland, United States.

OECD (2018), "Economic Assessments of the Benefits of Regulating Mercury – a Review", Working Party on Integrating Environmental and Economic Policies, Joint Meeting of the Chemicals Committee and the Working Party on Chemicals, Pesticides and Biotechnology, 23 January 2018, ENV/EPOC/WPIEEP/JM(2017)1/REV1.

OECD (2016), "Draft Agenda and Issues Paper - Meeting of the Environment Policy Committee (EPOC) at Ministerial Level", paper presented to EPOC on 28-29 September 2016, ENV/EPOC(2016)14/REV1.

OECD (2014), "OECD Expert Workshop on Economy-wide Nitrogen Balances and Indicators - Summary of Discussion Outcomes", ENV/EPOC/WPEI(2014)5.

OECD (2013), "Economy-wide Nitrogen Balances and Indicators - Concept and Methodology", ENV/EPOC/WPEI(2012)4/REV1.

OECD (2008), "An OECD Framework for Effective and Efficient Environmental Policies", paper prepared for the Meeting of the Environment Policy Committee (EPOC) at Ministerial Level, 28-29 April 2008, www.oecd.org/env/tools-evaluation/41644480.pdf.

OECD/ECLAC (2016), *OECD Environmental Performance Reviews: Chile 2016*, OECD Publishing, Paris, doi.org/10.1787/9789264252615-en.

PCE (2016), *Climate Change and Agriculture, Understanding the Biological Greenhouse Gases*, October 2016, Parliamentary Commissioner for the Environment, Wellington New Zealand, www.pce.parliament.nz/media/1678/climate-change-and-agriculture-web.pdf.

Ransom, K. M. et al. (2017), "A Hybrid Machine Learning Model to Predict and Visualize Nitrate Concentration throughout the Central Valley Aquifer, California, USA", *Science of the Total Environment*, 601–602.

Roberts, M. and A. Kolosseus (2014), "Science Sheds Light on Puget Sound Dissolved Oxygen", Department of Ecology, State of Washington, ecologywa.blogspot.fr/2014/03/new-science-sheds-light-on-puget-sound.html.

Rockström, J. et al. (2009), "A Safe Operating Space for Humanity", *Nature*, 461.

Rogelj, J. et al.; (2015), "Impact of Short-lived non-CO_2 Mitigation on Carbon Budgets for Stabilizing Global Warming", *Environmental Research Letters*, 10(7).

Salomon, M. et al. (2016), "Towards an Integrated Nitrogen Strategy for Germany", *Environmental Science & Policy*, 55 (2016).

Saunders P. (2010), "Dealing with Uncertainty, Precaution versus Science – Polar Opposites or a Continuum?", Proceedings of an OECD Workshop on the Economic and Trade Implications of Policy Responses to Societal Concerns in Food and Agriculture, 2-3 November 2009.

Seitzinger, S.P. and L. Phillips (2017), "Nitrogen Stewardship in the Anthropocene, How can Nitrogen Emissions be Reduced and Reused to Reduce Pressure on Ecosystems?", *Science*, 357(6349), science.sciencemag.org/content/sci/357/6349/350.full.pdf.

Socolow, R. (2016), "Fitting on the Earth: Challenges of Carbon and Nitrogen Cycle to Preserve the Habitability of the Planet", *Engineering*, 2(1).

Socolow, R. (1999), "Nitrogen Management and the Future of Food: Lessons from the Management of Energy and Carbon", *Proc Natl Acad Sci USA*;96(11).

Steffen, W. K. et al. (2015), "Planetary Boundaries: Guiding Human Development on a Changing Planet", *Science*, 347 (6223).

Sutton, M.A. et al. (2013), *Our Nutrient World: The Challenge to Produce More Food and Energy with Less Pollution*, Global Overview of Nutrient Management, Centre of Ecology and Hydrology, Edinburgh on behalf of the Global Partnership on Nutrient Management and the International Nitrogen Initiative.

Sutton, M.A. et al. (2011), *The European Nitrogen Assessment: Sources, Effects and Policy Perspectives*, Cambridge University Press.

Tomich, Th. P. et al. (2016), *The California Nitrogen Assessment: Challenges and Solutions for People, Agriculture, and the Environment*, University of California Press.

UNGA (1992), "Report of the United Nations Conference on Environment and Development", United Nations General Assembly, Rio de Janeiro, 3-14 June 1992, A/CONF.151/26 (Vol. I). 12 August 1992.

USEPA-SAB (2011), *Reactive Nitrogen in the United States: An Analysis of Inputs, Flows, Consequences and Management Options*, U.S. Environmental Protection Agency's Science Advisory Board, EPA-SAB-11-013, USEPA, Washington D.C., yosemite.epa.gov/sab/sabproduct.nsf/WebBOARD/INCFullReport/$File/Final%20INC%20Report_8_19_11(without%20signatures).pdf.

Vitousek, P.M. et al. (2009), "Nutrient Imbalances in Agricultural Development", *Science*, 324 (5934).

Walker, B. D et al. (2016), "Pacific Carbon Cycling Constrained by Organic Matter Size, Age and Composition Relationships", *Nature Geoscience*, 9, December 2016, doi.org/10.1038/ngeo2830.

Chapter 3. Examples of impact-pathway analysis and its translation into policy-making

This chapter provides examples of impact-pathway analysis to improve risk management of air pollution and water pollution. The examples illustrate the management of urban air pollution (Paris) and nitrogen deposition on forest ecosystems (in Germany). Other examples are the management of dead zones in Chesapeake Bay (eastern United States), the risk of lake pollution (in New Zealand) and groundwater pollution (western United States).

Measures have been taken to address nitrogen impacts on ecosystems and human health. Regarding atmospheric nitrogen emissions, OECD countries have adopted measures to reduce long-range transported emissions as well as targeted measures to reduce local emissions (e.g. in cities and to protect sensitive ecosystems). Similarly, measures have been taken to reduce nitrogen discharges into water, as well as targeted measures to reduce local emissions (e.g. at the watershed level and to protect vulnerable aquifers, lakes, estuaries and coastal waters). As a result, nitrogen emissions have been reduced in the OECD area over the last three decades. Emissions of nitrogen oxides (NO_x) have been reduced from both the stationary combustion installations in the energy sector and the transport sector. There have been reductions in emissions of ammonia (NH_3) from the agricultural sector. Emissions of nitrous oxide (N_2O) have also been reduced. Nitrogen inputs into surface waters have declined, both from point sources (municipal waste water treatment plants and industry) and from diffuse sources (as measured by the OECD national agricultural nitrogen balance).

However, while OECD nitrogen emissions have been reduced, the concentrations of key nitrogen forms in air, soil and water still remain much too high overall (Salomon et al., 2016). In particular, air quality standards with regard to nitrogen dioxide (NO_2) and particulate matter (PM) are still being exceeded regularly, particularly around busy roads. Terrestrial ecosystems are still being affected by eutrophication and, to a lower extent, by acidification.[1] Although emissions of N_2O decreased, there is no trend reversal of atmospheric N_2O concentration, which is still increasing (N_2O being a long-lived pollutant). Groundwater bodies and surface waters are still highly affected by nitrogen inputs.[2] Overall, there is little indication of any fundamental improvement in the eutrophication situation in marine waters.[3]

The disconnect between the reduction of nitrogen emissions and the persistence of impacts partly reflects a lag time between the first and its effects on the second. But this is not the only factor. The examples presented in this Chapter show that Impact-Pathway Analysis (IPA) can help manage impacts more cost-effectively by fostering evidence-informed policy making. Examples highlight the role of deposition analysis in urban smog control and critical load control for terrestrial ecosystems, as well as the role of leaching analysis in managing lake pollution. Two examples in the United States illustrate the role of IPA in managing two water pollution risks, eutrophication of a coastal zone (Chesapeake Bay) and nitrate (NO_3^-) contamination of an aquifer (Willamette Basin).

3.1 Case study 1: Impact-Pathway Analysis (IPA) and air pollution

3.1.1 Urban air pollution

In March 2014, Paris suffered a major peak of particle pollution that lasted for ten days (Figure 3.1). IPA revealed that half of coarse particles (PM_{10}) were ammonium nitrate (NH_4NO_3) particles formed by a combination of NO_x emitted mainly by urban transport and NH_3 originating from farming activity in relatively distant geographical areas (the northwest part of France and beyond) (PRIMEQUAL, 2015). This combination occurred because of the high pressure (anticyclonic) conditions that prevailed these days, which maintained high concentrations of NO_x and NH_3 (thermal inversion effect). The other half originated from combustion of biomass (wood heating), fuel combustion (including transport), volatile organic compounds (VOCs) and ammonium sulphate. This finding was made possible thanks to scientific progress in recent years, which has made the measurement of NH_4NO_3 more accurate.

Figure 3.1. Particle alert threshold exceeded in Paris in March 2014

On the left, the Eiffel Tower before the particle pollution episode. On the right, a photo taken at the same place on 14 March 2014.
Source: http://www.natura-sciences.com/environnement/particules-fines-pics-pollution810.html, accessed 27 June 2018.

This IPA finding had a direct implication for policy. It led the French authorities to argue that it was just as justified to act on fertiliser application practices as it was to act on traffic in order to curb urban air pollution. Measures taken in air pollutant emission zones included: setting speed limits on roads, making residential parking free of charge, calling on farmers and firms to temporarily limit fertiliser use and industrial activity (respectively), and promoting the use of public transport. In addition, at the height of the peak of pollution (17 March 2014) more stringent measures were taken, including: restricting vehicle use, reducing speed limits, restricting heavy truck traffic, making public transport free of charge, prohibiting the burning of green waste (including agricultural). This example shows how IPA led policy makers to not only address pollution emanating from the household heating and transport sectors, but also – and this is more unusual – from agriculture, which had historically not been considered when thinking about reducing urban air pollution.

The role of farming in urban air pollution has also recently been highlighted in the United Kingdom. IPA revealed that the particularly severe episode of smog experienced throughout the United Kingdom (from Cornwall to Aberdeen) between 26 March and 8 April 2014 was mainly driven by NH_4NO_3 particles resulting from agricultural NH_3 emissions in continental Europe, and not – as had been wrongly claimed – by Saharan dust (Vieno et al., 2016). Thunis et al., 2017 show that agricultural emissions have a significant impact on air quality in many EU cities.

3.1.2 Eutrophication of terrestrial ecosystems

In the German Land of Baden-Württemberg (BW) on-the-ground monitoring combined with high-resolution IPA revealed high nitrogen deposition loads on major parts of the Land, which chemical transport modelling and lower-resolution IPA had not revealed (Figure 3.2). For example, the annual average deposition on coniferous forests was found to be 47% higher than previously thought. The high-resolution IPA also revealed differences in habitat vulnerability to nitrogen deposition.

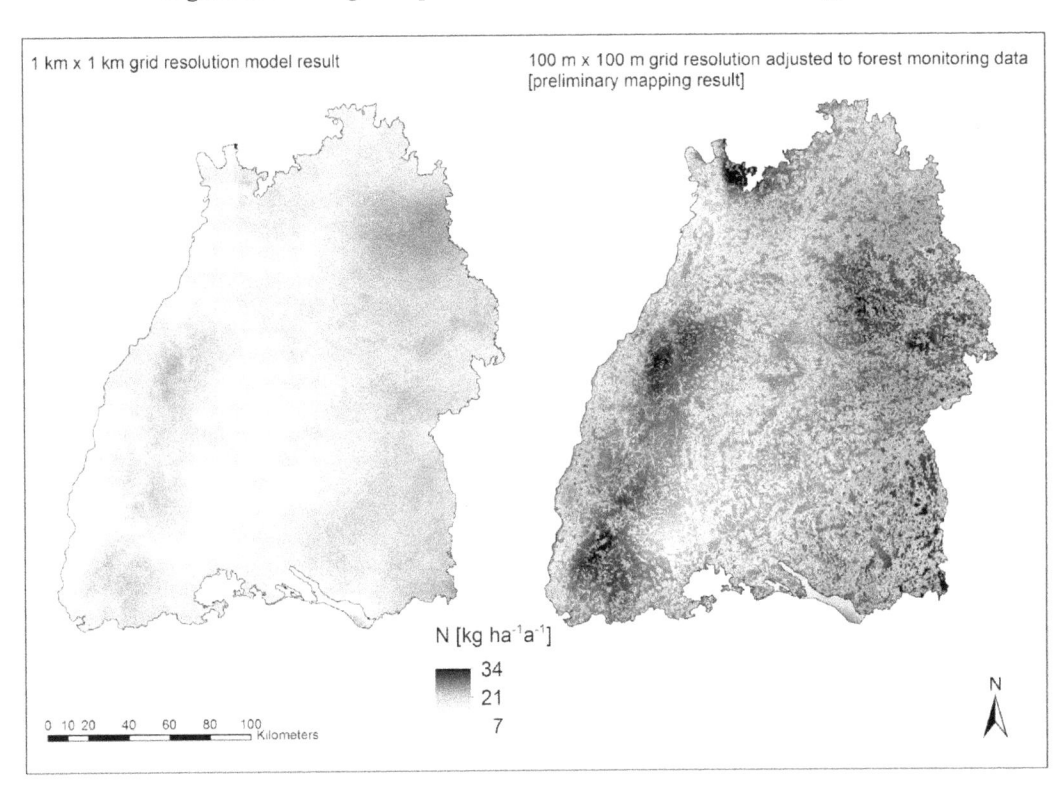

Figure 3.2. Nitrogen deposition loads in Baden-Württemberg, 2009

Note: Deposition of oxidised nitrogen and reduced nitrogen. Beyond the level of resolution, the significant difference between the two maps reflects a difference between the methods used to estimate nitrogen deposition, namely mapping adjusted to actual monitoring on the ground (right map) compared to modelling (left map). It should be noted that the resolution level strongly influences the estimation of critical loads for eutrophication and their exceedances, which are based on the dominant ecosystem in the grid area and the relevant EU legislation, namely the Habitat Directive for the higher resolution (right map) and the National Emission Ceilings Directive for the lower resolution (left map).
Source: LUBW (2016, 2018), Gauger (2017).

Such IPA finding has direct policy implications as it suggests the need for more effective regulation of nitrogen sources in emission zones. Today, emission threshold values are only imposed on traffic, industrial and large livestock facilities (which account for only 1% of agricultural nitrogen emissions) when critical loads for eutrophication are exceeded. As is the case throughout Germany, there is no trigger of threshold values for the other farming activities but an annual requirement put on all farms to estimate a farm-gate nitrogen balance. Yet, nitrogen deposition in BW mostly (55%) arises from agriculture, with transport and industry accounting for 22% and 23%, respectively.

IPA led the Netherlands to recently introduce site-specific regulations as part of a preventive approach to the protection of natural ecosystems. In this country, from 1 July 2015, the permitting of new economic activities (e.g. agriculture, industry, traffic) has been conditional upon prior assessment of the impact of nitrogen deposition on Natura 2000 areas (as estimated by the AERIUS calculation tool[4]) as part of a so-called Integrated Approach to Nitrogen (PAS). In the Netherlands, most (118 out of 160) Natura 2000 sites are affected by excess nitrogen deposition (Ministry of Economic Affairs, 2015).

Portugal is also moving toward better deposition risk assessment, as shown by recent studies on assessing Mediterranean ecosystem vulnerability to NH_3 (Pinho et al., 2016). However, it is necessary to define a workable grid level. No less than 231 habitat types are listed in the EU Habitat Directive (92/43/EC), each of which is likely to have its own critical load.

A further IPA refinement to address nitrogen deposition impacts on terrestrial biodiversity would be to assess critical nitrogen concentrations in soils of protected natural areas (in the soil solution). Indeed, historical loading characterises the soil species composition significantly. Soil bacteria (which play a key role in the nitrogen cycle) as well as mycorrhizal fungi (which contribute to plant growth) are very sensitive to changes in soil nitrogen ratios.

3.1.3 Policy relevance of IPA for air pollution risk management

Nitrogen dioxide (NO_2)

Despite a steady decline in NO_x emissions since 2000, most EU countries have at least one city where the annual average concentration of NO_2 exceeds (sometimes considerably) the EU's legal limit values (equal to the World Health Organisation (WHO) Air Quality Guideline) (Figure 3.3).[5] In 2013, some monitoring stations in France, Germany and the United Kingdom recorded annual average concentrations above twice the European Union (EU) limit value (Figure 3.3). In February 2017, the European Commission sent final warnings to France, Germany, Italy, Spain and the United Kingdom for failing to address repeated breaches of air pollution limits for NO_2.[6]

Figure 3.3. Nitrogen dioxide (NO₂) exceeds legal limits in many EU cities despite the reduction of nitrogen oxides (NOₓ) emissions at national level

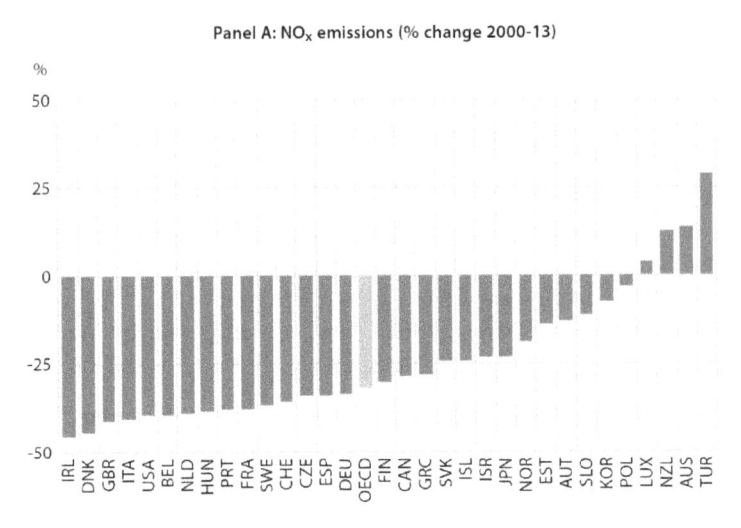

Panel A: NOₓ emissions (% change 2000-13)

Panel B: NO₂ concentrations (2013)

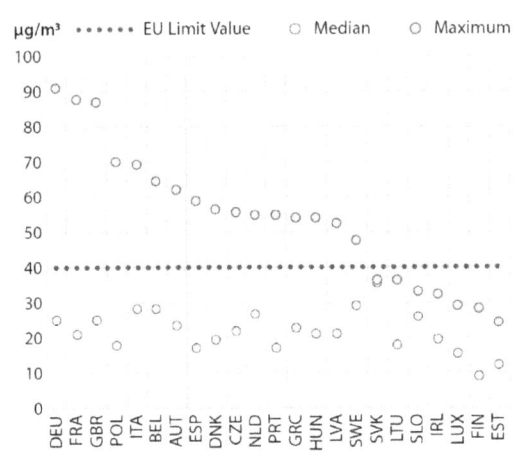

Note: NO₂ concentrations: average values recorded by monitoring stations.
Source: OECD (2015 for Panel A; 2017 for Panel B).

The repeated violations of air pollution limits for NO₂ are partly due to an underestimation of NOₓ emissions by vehicles. It appears that traffic contributes more to NOₓ emissions than previously thought, up to four times more (Karl et al., 2017).[7]

Nevertheless, local NO₂ concentrations are also highly dependent on the location of NOₓ emission sources and their atmospheric pathways. Increased use of IPA would better correlate NO₂ risk areas with NOₓ emission zones, thus facilitating policy decision.

Nitrogen aerosols

A similar observation can be made for PM pollution, for which NOₓ and NH₃ are precursors. Despite improvements, the populations of most EU countries remain chronically exposed to harmful levels of PM.[8] Fewer than one in three OECD countries comply with the WHO Air Quality Guideline for annual average fine particles (PM$_{2.5}$) exposure of 10 micrograms per m^3 (OECD, 2017). Outside the OECD area, exposure to PM$_{2.5}$ exposure in China and India continued to increase despite already extremely high levels (Figure 3.4). The number of premature deaths from PM$_{2.5}$ has been increasing between 2000 and 2015, both in emerging countries and OECD countries as a whole (Roy and Braathen, 2017).

Figure 3.4. Particulate matter (PM) concentrations exceed and are projected to continue to exceed legal limits in a business as usual scenario

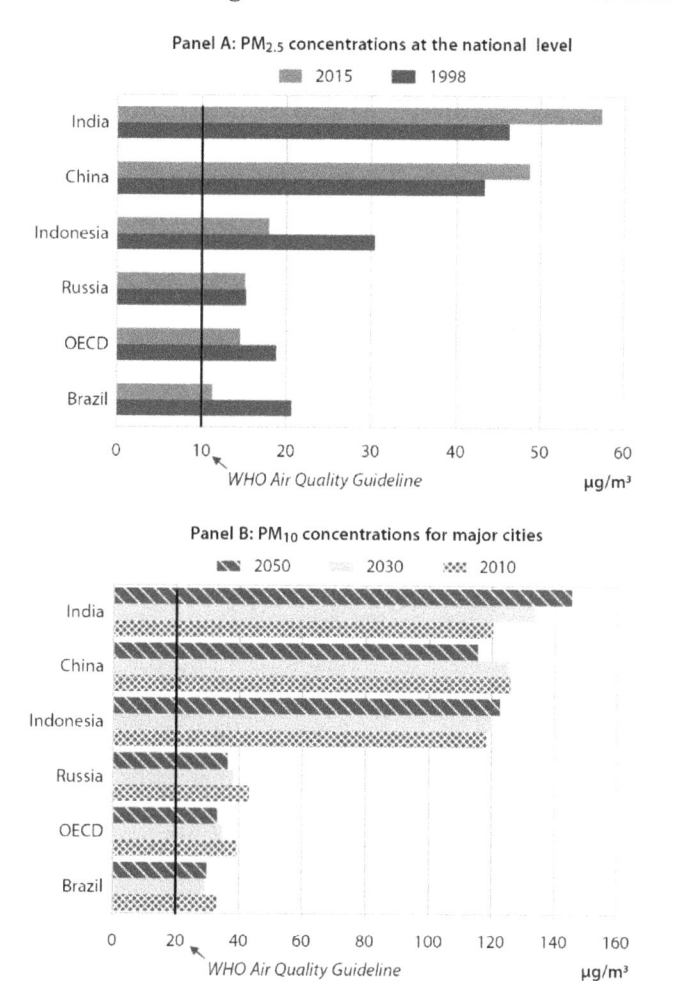

Panel A: PM$_{2.5}$ concentrations at the national level

Panel B: PM$_{10}$ concentrations for major cities

Panel A: Average concentrations of fine particles (PM$_{2.5}$) are derived from satellite observations, chemical transport models and monitoring stations.
Panel B: Annual average concentrations of coarse particles (PM$_{10}$) for cities with population over 100 000.
Source: OECD (2017 for Panel A; 2012 for Panel B).

The US Environmental Protection Agency's Science Advisory Board (USEPA-SAB) assessment mapped the risk of nitrogen aerosol deposition in the United States. As evidenced by the assessment, it can be expected than ammonium (NH$_4^+$) wet deposition occurs near or downwind of major agricultural centres, and that nitrate (NO$_3^-$) levels in wet deposition are consistent with NO$_x$ emissions[9] (USEPA-SAB, 2011). Increased use of IPA would better correlate nitrogen aerosol risk areas with NO_x and NH_3 emission zones, thus facilitating policy decision.

Ground-level ozone (GLO)

A similar observation can be made for GLO pollution, for which NO_x is a precursor. Despite improvements, the populations of most EU countries remain chronically exposed to harmful levels of GLO (Figure 3.5).[10] Reduction of NO_x emissions has not (linearly) translated into GLO reduction. Instead, GLO trends over recent decades show complex patterns (Cooper et al., 2014).

As with NO_2, changes in GLO concentrations will depend not only on the level of precursor emissions, but also on the location of emissions, atmospheric pathways and climatic conditions.[11] In particular, GLO concentrations are often higher downwind of urban areas than in urban areas themselves.[12]

3. EXAMPLES OF IMPACT-PATHWAY ANALYSIS AND ITS TRANSLATION INTO POLICY-MAKING | 79

Figure 3.5. Urban population exposure to ground-level ozone (GLO) is and is projected to remain a concern in a business as usual scenario

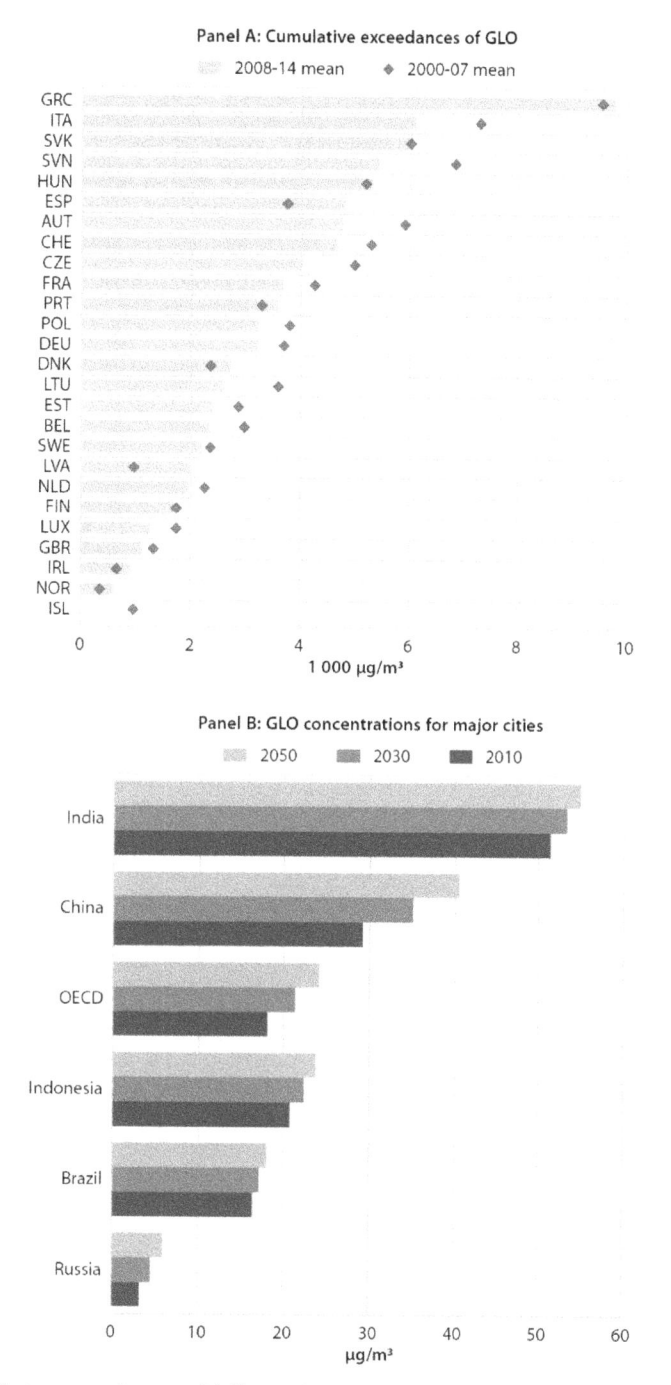

Panel A: Cumulative exceedances of GLO

Panel B: GLO concentrations for major cities

Panel A: Cumulative exceedances of daily maximum 8-hour mean exposures above 70 μg/m³ for all days in a year, based on measurements at ground stations, selected European countries. By comparison, World Health Organisation (WHO) Air Quality Guideline and European Union (EU) target values are, respectively, 100 μg/m³ and 120 μg/m³ for maximum daily 8-hour mean exposure.
Panel B: Annual average GLO concentrations for cities with population over 100 000.
Source: OECD (2017 for Panel A; 2012 for Panel B).

3.2 Case study 2: Impact-Pathway Analysis (IPA) and water pollution

3.2.1 Coastal water pollution

The Chesapeake Bay Watershed covers 90 000 square miles (23 million hectares) across six U.S. States (Delaware, Maryland, Pennsylvania, New York, Virginia, West Virginia) and Washington D.C. Land use is forest (64%), agriculture (24%), urban (8%) and other (4%). Nitrogen, phosphorus and sediment loads delivered from the watershed to the tidal waters of the Bay are the primary concern. They translate into low to no dissolved oxygen (DO) in the Bay and tidal rivers every summer. Required DO levels have been established for the different species living in the Bay as a function of the depth of the water (Figure 3.6).

Figure 3.6. Chesapeake Bay's criteria for dissolved oxygen

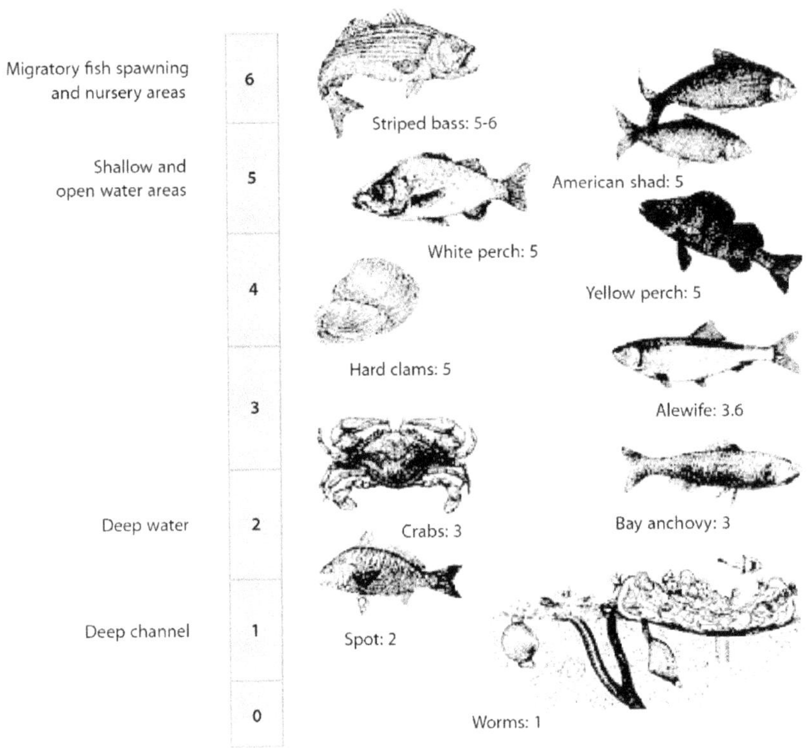

Depth zone	Scale	Species: DO (mg/l)
Migratory fish spawning and nursery areas	6	Striped bass: 5-6
Shallow and open water areas	5	American shad: 5; White perch: 5; Yellow perch: 5; Hard clams: 5
	4	
	3	Alewife: 3.6
Deep water	2	Crabs: 3; Bay anchovy: 3
Deep channel	1	Spot: 2
	0	Worms: 1

Note: The scale from 1 to 6 is based on the minimum amount of oxygen (mg/l) required for the survival of species.
Source: Linker et al. (2016).

In December 2010, a TMDL of nitrogen and phosphorus allowed to enter the tidal Bay has been set to comply with DO standards. The TMDL covers major land-based sources of nutrients in the watershed (agriculture, sewage) and atmospheric deposition of nitrogen in the watershed. It is the first TMDL scheme in the United States that takes into account atmospheric deposition of nitrogen. The projected reduction in NO_x emissions at the national level (pursuant to the Clean Air Act) is deducted from the TMDL, thereby reducing the nutrient reduction effort required from land-based sources. Another TMDL has been established for direct atmospheric deposition of nitrogen in tidal waters.

The relevant "airshed" – the area where emission sources contribute most to deposition in the Bay's watershed – was delineated for each nitrogen form to model nitrogen deposition. The NH_3 airshed is similar to the NO_x airshed, but slightly smaller (Figure 3.7). Both airsheds are approximately nine times larger than the Bay watershed. Approximately 50% of nitrogen deposits in the Bay come from sources located in the Bay watershed. Another 25% comes from the remaining area in the Bay airshed. The final 25% comes from net ocean exchange.

Figure 3.7. Chesapeake Bay airsheds for nitrogen oxides (NO_x) and ammonia (NH_3)

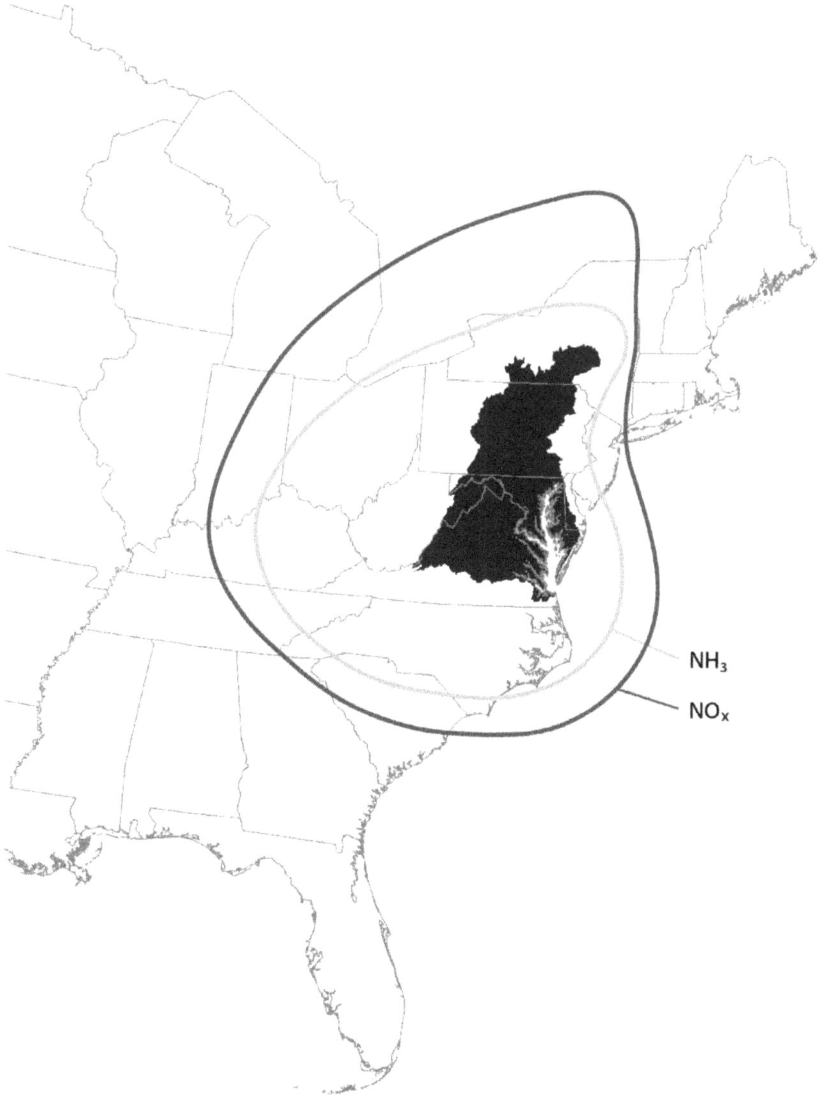

Note: The highlighted area within the airsheds represents the area of the Chesapeake Bay watershed.
Source: Linker et al. (2016).

Integrated modelling supports IPA analysis. It involves a step-by-step approach. First, data from a land use change model, an airshed model and a "scenario builder" software are transmitted to a watershed model (Figure 3.8). The Land Use Change model predicts changes in land use, sewerage, and septic systems given changes in land use policy. The Airshed Model, a national application of Community Multiscale Air Quality Model (CMAQ), predicts changes in deposition of inorganic nitrogen due to changes in emissions.[13] The "scenario builder" software combines the output of these models with other data sources, such as the U.S. census of agriculture, to generate inputs to the Watershed Model.

Figure 3.8. Chesapeake Bay Programme modelling framework

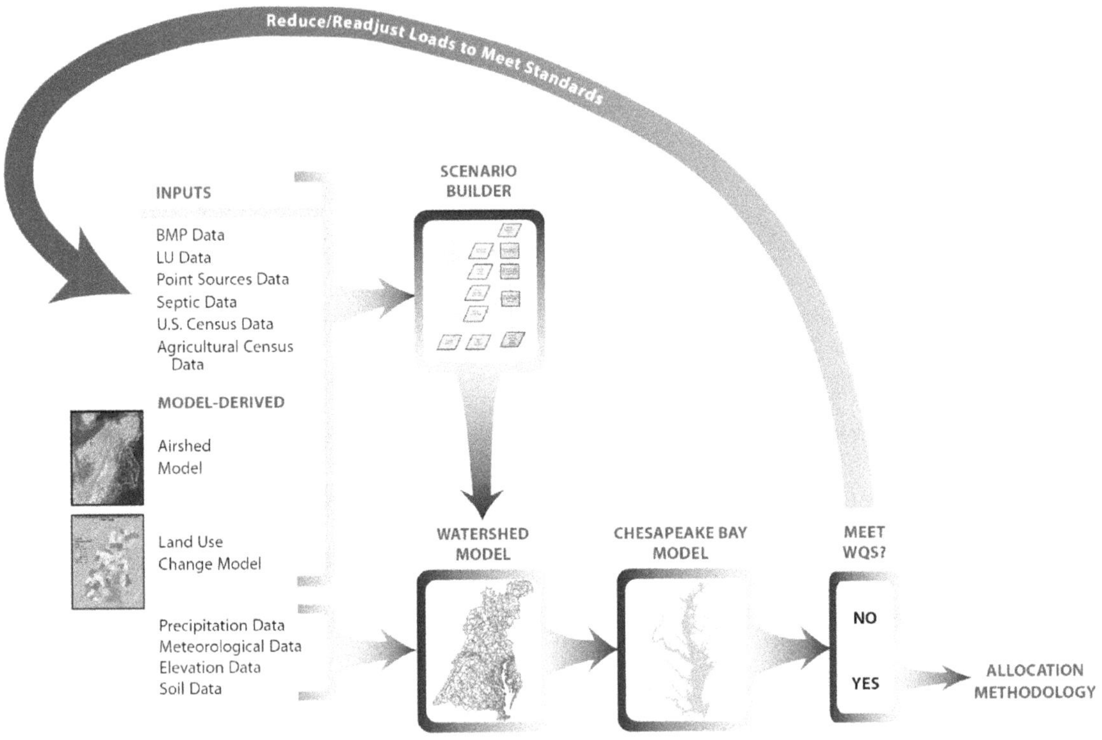

Source: www.chesapeakebay.net/who/group/modeling_team, accessed 11 April 2018.

The Watershed Model then predicts the loads of nitrogen, phosphorus, and sediment that result from the given inputs.[14] The estuarine Water Quality and Sediment Transport Model (WQSTM) (also known as the Chesapeake Bay Model) predicts changes in Bay water quality due to the changes in input loads provided by the Watershed Model.[15] As a final step, a water quality standard analysis system examines model estimates of DO, chlorophyll, and water clarity to assess in time and space the attainment of the Bay water quality standards

Direct deposition of nitrogen in tidal waters is estimated on the basis of reductions at the federal level of mobile emissions and pursuant to the Cross-State Air Pollution Rule[16] plus reductions at the State level.[17] NO_x deposits on tidal waters have declined since the mid-1980s while NH_3 deposits are stable or increasing (Table 3.1). The TMDL target was set at 15.7 million pounds (7 100 tonnes) of N by 2020, representing a 40% reduction in nitrogen deposition compared to 1985.

Table 3.1. Estimated direct deposition of nitrogen to the tidal Chesapeake Bay

(Millions of pounds N)

Scenario	NO$_x$ wet	NO$_x$ dry	NH$_3$ wet	NH$_3$ dry	Total inorganic nitrogen	Wet organic nitrogen	Total nitrogen	% change from 1985
1985	6.6	13.1	3.3	2.0	25.0	1.0	26.1	-
2002	4.8	10.0	3.6	2.1	20.5	1.0	21.6	17
2010	3.3	6.8	3.5	2.8	16.4	1.0	17.4	33
2020 target	2.6	5.1	3.7	3.2	14.6	1.0	15.7	40
2030 target	2.2	4.3	4.0	4.1	14.6	1.0	15.6	40

Source: Linker et al. (2016).

The combined management of atmospheric and land-based nitrogen inputs in Chesapeake Bay reflects the reality of the nitrogen cycle. Such an IPA makes the risk management of hypoxia in the bay more cost effective. The Chesapeake Bay Programme has been effective as a whole since the TMDL was introduced, although progress remains to be made to meet the nitrogen load target of 2025 (Table 3.2). Estimated nitrogen loads in the Bay watershed decreased by 9% between 2009 and 2016. Agriculture accounts for 42% of the remaining nitrogen loads, followed by discharges from urban runoff and sewage (33%) and atmospheric deposition (25%) (Table 3.2).

Table 3.2. Estimated nitrogen loads in the Chesapeake Bay

(Millions of pounds N)

Sector	2009	2016	% change 2009-16	2016%	2025 target
Agriculture	113.8	108.0	-5	42	71.9
Urban runoff	39.7	41.4	4	16	28.8
Sewage and combined sewer overflow	52.2	36.4	-30	14	37.9
Septic tanks	8.4	8.7	3	3	6.3
Forest and non-tidal water deposition	46.2	45.6	-1	18	47.1
Deposition to watershed	3.1	1.3	-56	1	-
Deposition to tidal water	19.4	16.6	-15	6	15.2
Total	282.7	258.1	-9	100	192.4[1]

1. Excluding deposition to tidal water.
Source: www.chesapeakebay.net/indicators/indicator/reducing_nitrogen_pollution, accessed 22 September 2017.

More generally, models that predict how nitrogen from the air is deposited in the sea could be useful for managing the risk of algal blooms. For example, researchers have simulated nitrogen deposition in the North Sea and suggested that by superimposing weather forecast data, it would be possible to predict algal blooms (Djambazov and Pericleous, 2015).

3.2.2 Lake water pollution

As nitrogen moves into groundwater to a lake, leaching from different parts of the lake basin reaches the lake at different times and the damage (e.g. eutrophication) is temporally differentiated. Like spatial differentiation in the previous example

(Chesapeake Bay), a policy that incorporates time differences is likely to be more cost-effective than a policy that does not. This is what recent studies have sought to demonstrate.

Cox et al., 2013 undertook a simplified analysis of agricultural nitrogen transport pathways through the groundwater system to manage NO_3^- pollution risk of Lake Rotorua. A geophysical model was used to estimate mean groundwater residence times at the parcel level in the lake basin (Figure 3.9). The authors suggest that time-specific regulation is cost-effective in watersheds where nitrogen transport to the lake is primarily through groundwater (and not runoff) and where transport times are highly differentiated.

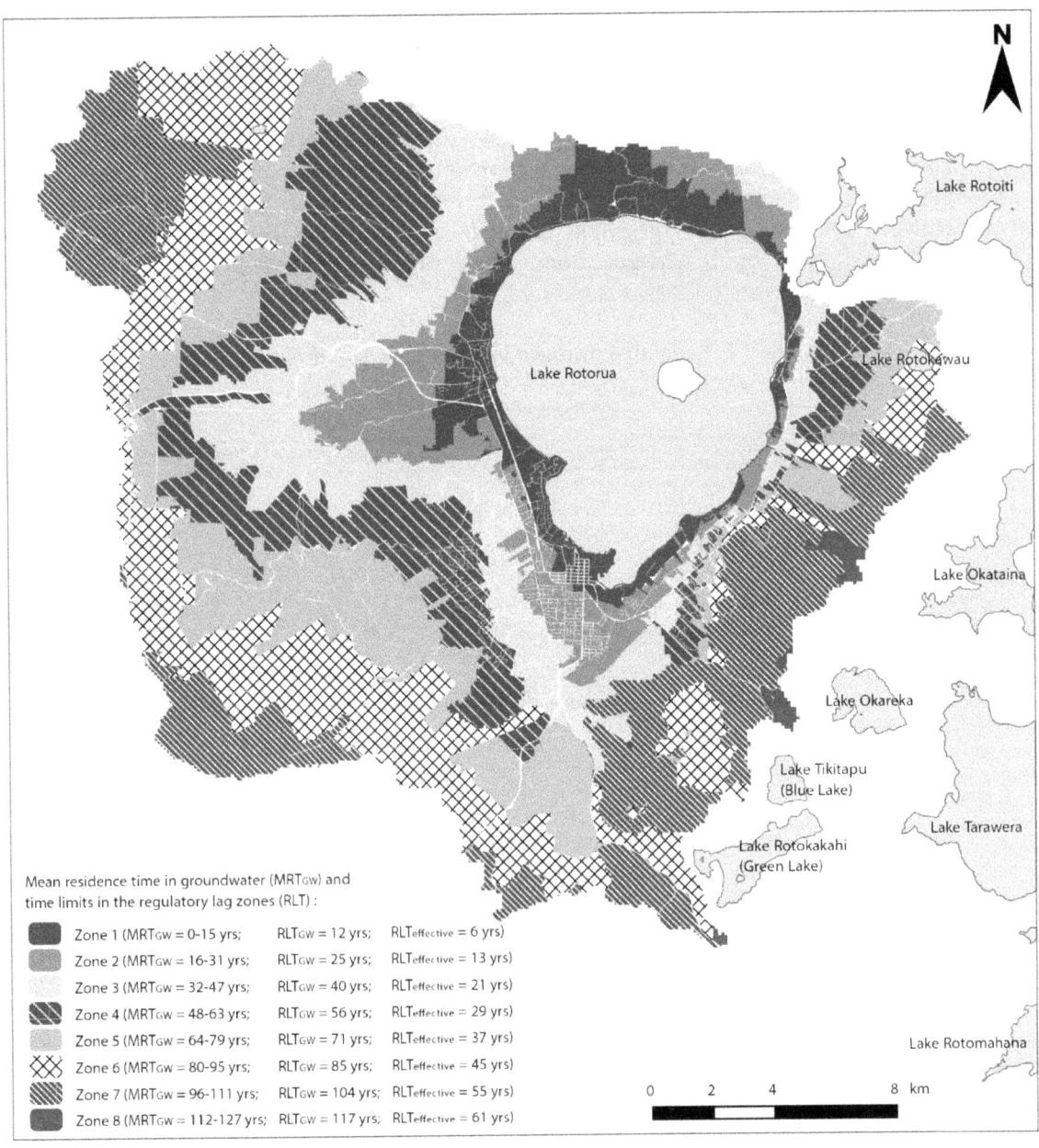

Figure 3.9. Nitrogen transport times in the Rotorua Lake basin

Source: Cox et al. (2013).

"Vintage Nitrogen Trading" (VNT) can be a cost-effective eutrophication risk management tool for lakes for which IPA can predict nitrogen inputs over time (in part, following Anastasiadis et al., 2013). In a vintage trading scheme, the regulator provides a supply of allowances for each vintage year, where the vintage year corresponds to the year when nitrogen will arrive in the lake. Allowances therefore represent rights to contribute to lake loads in a particular year which equate to conditional rights to leach nitrogen from farms[18] depending on groundwater lag time. Under regulation, farmers must surrender allowances each year to cover the lake loads that will be caused some time in the future by the nitrogen lost from their property that year.

Under such VNT scheme, each year farmers would need to match their leaching with allowances from the vintage that corresponds to the current year plus their lag time. For example, suppose a farmer with a lag time of 30 years leaches 100 kg of nitrogen in 2018; he would need, in 2018, surrender 100 kg of 2048 vintage allowances. Nitrogen also travels via soil surface runoff (quick-flow). Suppose 50% of nitrogen travels via runoff and 50% via groundwater. Then, the farmer would need, in 2018, surrender 50 kg of 2018 vintage allowances and 50 kg of 2048 vintage allowances.

Any catchment with groundwater lags will have a legacy load: nitrogen in the groundwater from historical leaching that is yet to be realised as lake loads. As it is very difficult to prevent nitrogen already in the groundwater from reaching the lake, regulators must account for legacy loads when setting environmental targets for all sources of nitrogen. This is frequently done by a gradual strengthening of the environmental target over time. Anastasiadis et al., 2011 discuss the design of regulation that allows for legacy loads.

The ability to respond (adaptability) to change is an important feasibility criterion for VNT systems, which require a long period of implementation. Like all tradable permit instruments, VNT systems offer great flexibility in how to reduce nitrogen and where (in the lake basin) to meet the cap. However, once implemented, it may be difficult to revise the cap down (for example, if new evidence on lake water quality emerges). This problem may to some extent be mitigated by the use of "phases" defined in advance, after which modifications of the cap can be introduced to be applied in the following phases (Drummond et al., 2015).

3.2.3 Groundwater contamination

In Oregon, United States, in 2004, the Southern Willamette Valley (SWV) was declared a groundwater management area (GWMA) due to the high concentration of NO_3^- that has been found in many private wells, with concentrations often exceeding the drinking water standard (10 mg/l). In fact, the SWV-GWMA - 230 square miles (almost 60 000 ha) - has not been delineated on a scientific basis (on the basis of an IPA) but according to criteria of administrative feasibility; it is the area in which public water systems abstract their drinking water.[19]

The administrative boundaries of the SWV-GWMA do not coincide with the risk area or the emission zone within the meaning of the IPA. It would be more appropriate to delineate the Willamette Lowland area (as a hydrogeological entity) – and not just a portion as is the case with SWV-GWMA – to manage the risk of groundwater contamination in the Willamette Basin (Box 3.1). Indeed, NO_3^- contamination is only for shallow groundwater - the shallow portion of the Willamette Aquifer adjacent to the Willamette River, known as the "alluvial aquifer".[20]

Instead, SWV-GWMA focuses on the protection of public water systems and it is assumed that each public water system has its own pathways of contamination. This led to the delineation of protection zones around each public well (the drinking water protection zones). "Source water assessments" identify potential sources of contamination in the drinking water protection zone (e.g. agriculture, septic tanks, abandoned wells, high density housing) and assign a level of risk (high, medium or low) based on general criteria developed by the USEPA.

This is a kind of IPA but on the wrong scale. Using the terminology of the IPA, the "risk area" is the alluvial aquifer and the "emission zone" is the Willamette Lowland area. Given the connectivity between the alluvial aquifer and the Willamette River, risk management focused on the alluvial aquifer would promote synergy between groundwater quality protection measures and surface water quality protection measures (such as TMDL).

Box 3.1. Groundwater recharge pathways in the Willamette Basin, Oregon

Conlon et al., 2005 estimated the recharge pathways of the Willamette Aquifer using the Precipitation-Runoff Modelling System (PRMS). The PRMS model shows high levels of groundwater recharge in the Coast Range, Western and High Cascade areas. However, it is likely that rainwater will only seep to a shallow depth and then flow into the streams of these areas. From the basin's point of view, this infiltration is not a recharge of the groundwater but a superficial flow in the soil zone. It follows that most of the precipitation in the Coast Range, Western, and High Cascade areas is discharged into streams in these areas and is not available as a source of groundwater in the lowlands (Figure 3.10). Consequently, shallow groundwater recharge in the lowland area occurs locally.

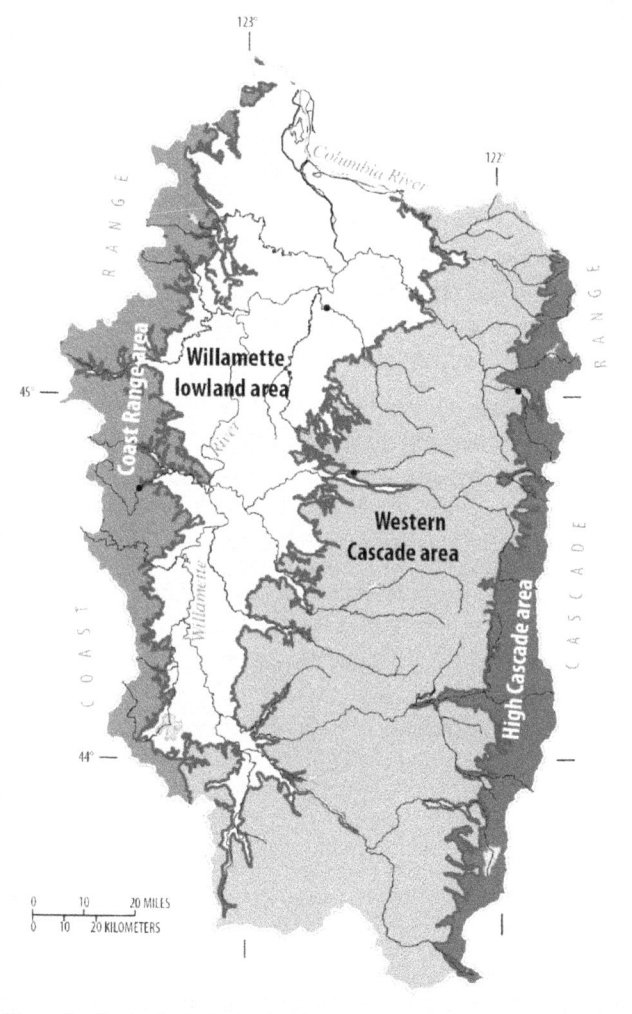

Figure 3.10. Hydrology of the Willamette Basin

Note: The Willamette Basin includes a lowland area between the Coast Range and the Cascade Range.
Source: Conlon et al. (2005).

3.2.4 Policy relevance of IPA for water pollution risk management

The example of "ocean dead zones" will serve to illustrate the policy relevance of IPA for water pollution risk management. Dead zones are steadily increasing, including in the OECD area, making the identification of the sources of nutrients that cause them a priority (Box 3.2).

Box 3.2. A compelling evidence of the rapid increase in ocean dead zones

As oxygen supply decreases in bottom waters, concentrations may fall below levels necessary to maintain animal life. This low oxygen condition is known as "hypoxia". Hypoxic areas are sometimes referred to as "dead zones", a term that was first applied to the hypoxic area of the northern Gulf of Mexico, which receives large amounts of nutrients from the Mississippi and Atchafalaya river basins (Rabalais et al., 2010). When fishermen trawl in these zones little to nothing is caught.

Ocean dead zones are closely associated with anthropogenic activities. The worldwide distribution of dead zones is related to major population centres and watersheds that export large quantities of nutrients.

The negative environmental effects of dead zones include loss of habitat, direct mortality, decreased food resources and altered migration for many fish species (bottom-dwelling and pelagic) (Breitburg et al., 2009). Increasing nutrient loads also affect composition of the phytoplankton (Turner et al., 1998). Dead zones also alter ecosystem services such as nutrient cycling (Sturdivant et al., 2012).

Dead zones vary according to seasons. In temperate latitudes, bottom waters can remain hypoxic for hours to months during summer and autumn. The good news is that coastal marine systems can be restored provided sustained efforts are undertaken to reduce nutrient loads. However, once a coastal marine system develops a dead zone, it often becomes an annual event (Baird et al., 2004).

Available data on dead zones provide compelling evidence of a rapid increase in the fertility of many coastal ecosystems since about 50 years ago. Until the late 1960s, there were scattered reports of dead zones in North America and Northern Europe (Diaz et al., 2013). Two decades later, in the early 1990s, dead zones had become prevalent in North America, Northern Europe and Japan (ibid). In the 2000s, dead zones extended their geographic reach to South America, Southern Europe and Australia (ibid).

Some 884 coastal areas around the world have been identified as experiencing the effects of eutrophication; of these, 576 have problems with hypoxia, 234 are at risk of hypoxia and, through nutrient management, 74 can be classified as recovering (Figure 3.11). This figure does not include the likely many unreported hypoxic areas in the tropics because of the lack of local scientific capacity for their detection (Altieri et al., 2017). It has been estimated that more than 10% of coral reefs are at high risk of hypoxia (ibid).

Ocean dead zones are particularly vulnerable to climate change: according to Altieri and Gedan 2015, almost all dead zones are in regions that will experience at least a 2°C temperature increase by the end of the century. Climate change exacerbates hypoxic conditions by increasing sea temperature, ocean acidification, sea level, precipitation, wind and storms.

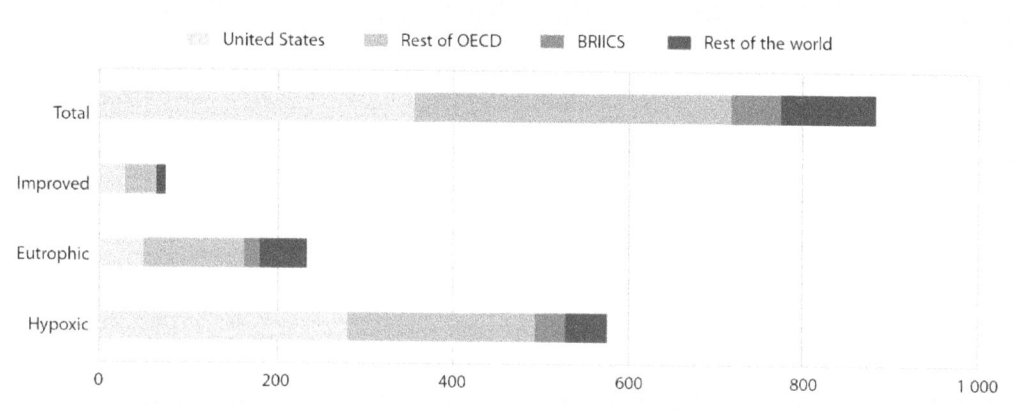

Figure 3.11. Global distribution of ocean dead zones

- Hypoxic - Areas with scientific evidence that hypoxia was caused, at least in part, by an overabundance of nitrogen and phosphorus. These areas are called "dead zones".
- Eutrophic - Areas that exhibit effects of eutrophication, including elevated nitrogen and phosphorus levels, elevated chlorophyll a levels, harmful algal blooms, changes in the benthic community, damage to coral reefs, and fish kills. These areas are at risk of developing hypoxia. Some of the areas may already be experiencing hypoxia, but lack conclusive scientific evidence of the condition.
- Improved - Areas that once exhibited low dissolved oxygen levels and hypoxia, but are now improving.
- BRIICS - Brazil, Russia, India, Indonesia, China and South Africa.

Source: Data collected by Robert Diaz, Virginia Institute of Marine Science, as of 22 June 2018.

UNEP, 2012 has correlated the formation of dead zones in the ocean with the growth of agricultural regions, cities and coastal development. According to UNEP, 2012, "the often inefficient use of fertiliser has resulted in releases of nitrogen and phosphorus into waterways and groundwater, which, together with nutrient losses from manure and inadequate sewage treatment has resulted in a substantial increase in nutrient discharges both directly in the coastal water and via rivers receiving emissions from upstream population centres and agriculture".

However, dead zones remain a concern even in the OECD area despite improvements in the efficiency of nitrogen use and improved nitrogen budgets in agriculture in the 2000s (Figure 3.12). As with the Chesapeake Bay mentioned above (see Section 3.2.1), the creation of dead zones is not only due to agriculture or land-based sources. An analysis of the different sources of export of river nitrogen to the sea shows that the sum of atmospheric deposition and natural biological fixation actually exceeds agricultural inputs, with the share of wastewater being the lowest (Table 3.3).[21] It is necessary to identify the different sources of nitrogen, their emission zones and the effective tools to manage the risk of dead zones they create, a role for the IPA.

Figure 3.12. OECD agricultural nitrogen: fewer surpluses and more efficiency in use

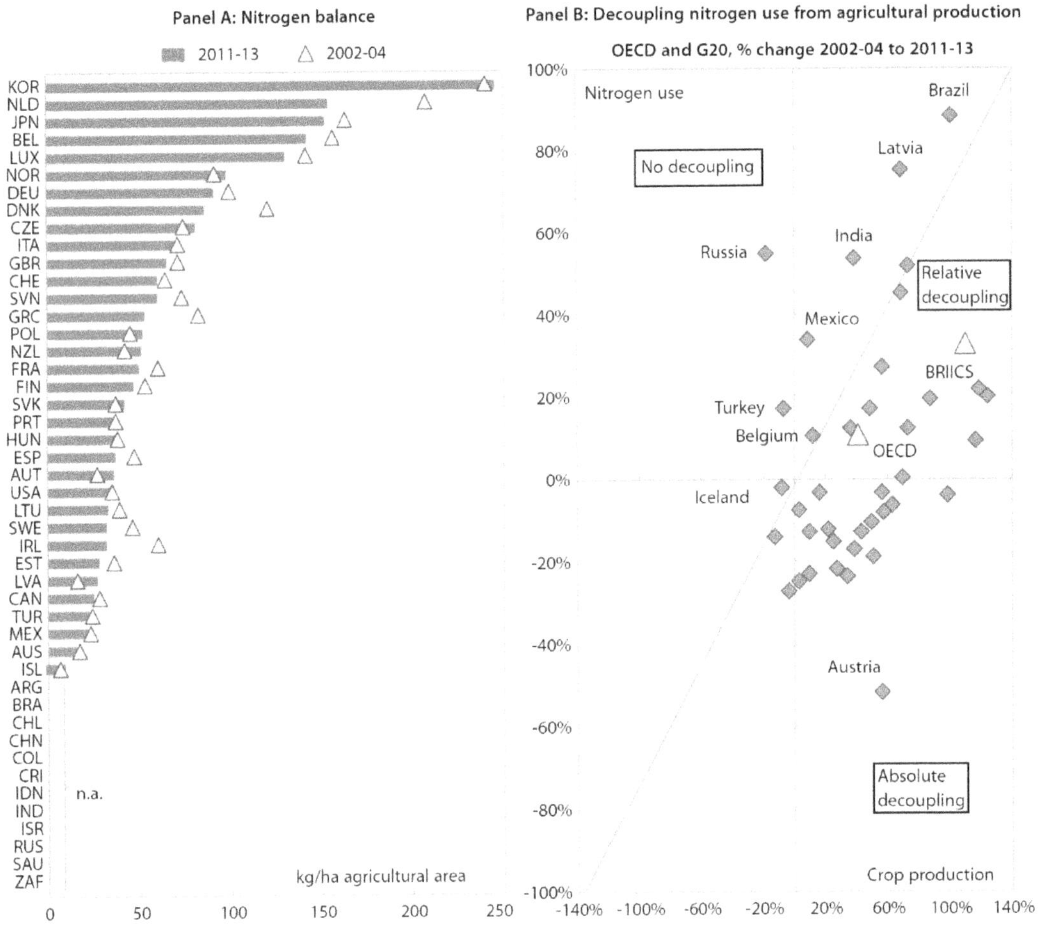

Panel A: National balance at soil surface.
Panel B: Consumption of commercial fertilisers in kg/ha of agricultural area. Crop production value in USD using 2010 prices and PPPs. OECD excludes the Czech Republic.
Source: OECD (2017).

Table 3.3. Sources of river nitrogen exports to coastal waters

(Million tonnes N)

Area	2000 Deposition and natural production[1]	2000 Agriculture[2]	2000 Sewage[3]	2000 Total	2030 Deposition and natural production[1]	2030 Agriculture[2]	2030 Sewage[3]	2030 Total	% change (total) 1970-2000	% change (total) 2000-30
OECD	6.4	4.4	1.8	12.7	5.7	4.3	2.0	12.0	10	-5
BRIC	11.9	8.7	1.4	21.9	9.0	12.9	2.4	24.3	57	11
RoW	12.7	5.0	1.0	18.6	10.8	6.5	1.6	18.9	26	2
World	30.9	18.0	4.1	53.1	25.4	23.7	6.0	55.2	33	4
World (%)	58	34	8	100	46	43	11	100		

BRIC (Brazil, Russia, India, China); RoW (Rest of the World, i.e. all countries except OECD and BRIC)
1. Nitrogen deposition and natural biological fixation (in non-cultivated areas)
2. Nitrogen surplus on cultivated areas
3. Nitrogen effluents from public sewerage.
Source: OECD (2008).

Notes

[1] In 2009, some 48% of Germany's natural and semi-natural terrestrial ecosystems were affected by eutrophication and 8% were affected by acidification (SRU, 2015).

[2] For example, many bodies of groundwater still fail to achieve good chemical status according to the EU Water Framework Directive (WFD) due to excessive nitrate (NO_3^-) concentrations (>50 mg/l). Similarly, many rivers still exceed the WFD good chemical status of 2.5 mg/l NO_3^- and very few meet the WFD good ecological status due to altered morphology and eutrophication.

[3] For example, eutrophication remains a problem for almost the entire Baltic Sea (HELCOM, 2013).

[4] www.aerius.nl/files/media/Publicaties/Documenten/aerius_the_calculation_tool_of_the_dutch_integrated_approach_to_nitrogen.pdf.

[5] NO_2 is directly harmful to human health (see Annex A).

[6] europa.eu/rapid/press-release_IP-17-238_en.htm.

[7] Until now NO_x emission levels were mainly calculated by collecting emission data at laboratory testing facilities and subsequently extrapolating them in models. However, the amount of pollutant emissions that vehicles emit on a daily basis depends on numerous factors, for example on individual driving behaviour.

[8] In 2014, respectively 16% and 8% of the EU-28 urban population were exposed to coarse particles (PM_{10}) and fine particles ($PM_{2.5}$) levels above the EU limit values; the proportions increase to 50% and 85% when considering the more stringent WHO Air Quality Guideline values (EEA, 2016).

[9] Although the decreases in deposition are probably not linearly proportional to the decreases in emissions. For example, a 50% reduction in NO_x emissions is expected to result in an approximate 35% reduction in NO_3^- concentration and deposition (USEPA-SAB, 2011).

[10] In 2014, 8% of the EU-28 urban population was exposed to GLO levels above the EU target value and 96% to levels higher than the more stringent WHO Air Quality Guideline value (EEA, 2016).

[11] GLO concentrations peak in summer. There is a large day-do-day variability reflecting meteorological conditions: GLO is highest under stagnant conditions associated with strong subsidence inversions.

[12] Near emission sources, NO_x reduces GLO by titration, while net GLO formation occurs some distance downwind of NO_x sources, depending on temperature and atmospheric dispersion.

[13] The Airshed Model combines a regression model of wet deposition with CMAQ estimates of dry deposition. CMAQ covers the North American continent on a grid of 36 km x 36 km; a finer grid (12 km x 12 km) is used on the Chesapeake Bay watershed.

[14] The Watershed Model was first developed in 1982; it is now in its fifth development phase.

[15] The WQSTM Model also tracks the transport of sediments, including their resuspension, by modelling the waves in the Bay estuary.

[16] The Cross-State Air Pollution Rule addresses air pollution from upwind states that crosses state lines and affects air quality in downwind states. It replaced the Clean Air Interstate Rule in 2015.

[17] State Implementation Plans to meet National Ambient Air Quality Standards.

[18] Assuming that the nitrogen loads in the lake come mainly from farmland.

[19] In Oregon, private wells are not subject to the laws protecting drinking water. Therefore, their owners are not required to adhere to drinking water standards (and are often unaware of contamination issues as NO_3^- cannot be tasted, seen or smelled).

[20] There is ample evidence linking high NO_3^- values with alluvium adjacent to the Willamette River in the 100-year floodplain.

[21] The relative importance of agricultural sources is projected to increase by 2030 in a business as usual scenario (Table 3.3).

References

Altieri, A.H. et al. (2017), "Tropical Dead Zones and Mass Mortalities on Coral Reefs", *Proc Natl Acad Sci U S A.*, 114(14).

Altieri, A.H. and K.B. Gedan (2015), "Climate Change and Dead Zones", *Global Change Biology*, 21.

Anastasiadis, S. et al. (2011), "Does Complex Hydrology Require Complex Water Quality Policy? NManager Simulations for Lake Rotorua", Working Paper 11-14, Motu Economic and Public Policy Research, Wellington, doi.org/10.2139/ssrn.1975569.

Anastasiadis, S. et al. (2013), "Does Complex Hydrology Require Complex Water Quality Policy?", *Australian Journal of Agricultural and Resource Economics*, 58(1), doi/10.1111/1467-8489.12024/epdf.

Baird, D. et al. (2004), "Consequences of Hypoxia on Estuarine Ecosystem Function: Energy Diversion from Consumers to Microbes", *Ecological Applications*, 14(3).

Breitburg, D.L. et al. (2009), « Hypoxia, Nitrogen, and Fisheries: Integrating Effects Across Local and Global Landscapes", *Annual Review of Marine Science*, 1.

Conlon, T.D. et al. (2005), *Ground-Water Hydrology of the Willamette Basin, Oregon*, U.S. Geological Survey, Scientific Investigations Report 2005–5168.

Cooper, O.R. et al. (2014), "Global Distribution and Trends of Tropospheric Ozone: An Observation-based Review", *Elementa: Science of the Anthropocene*, 2(29).

Cox, T.J. et al. (2013), "An Integrated Model for Simulating Nitrogen Trading in an Agricultural Catchment with Complex Hydrogeology", *Journal of Environmental Management*, 127 (2013), doi.org/10.1016/j.jenvman.2013.05.022.

Diaz, R.J. et al. (2013), "Hypoxia", in: Noone, K.J. et al. (eds.), *Managing Ocean Environments in a Changing Climate*, Elsevier, New York.

Díaz, R.J. and R. Rosenberg (2008), "Spreading Dead Zones and Consequences for Marine Ecosystems", *Science*, 321.

Djambazov, G. and K. Pericleous (2015), "Modelled Atmospheric Contribution to Nitrogen Eutrophication in the English Channel and the Southern North Sea", *Atmospheric Environment*, 102, do.org/10.1016/j.atmosenv.2014.11.071.

Drummond, P. et al. (2015), "Policy Instruments to Manage the Unwanted Release of Nitrogen into Ecosystems – Effectiveness, Cost-Efficiency and Feasibility", paper presented to the Working Party on Biodiversity, Water and Ecosystems at its meeting on 19-20 February 2015, ENV/EPOC/WPBWE(2015)8.

EEA (2016), *Air Quality in Europe — 2016 Report*, EEA Report N° 28/2016, European Environment Agency, Copenhagen, doi.org/10.2800/80982

Gauger, Th. (2017), "Reaktiver Stickstoff in der Atmosphäre von Baden-Württemberg - Interimskarten der Ammoniakkonzentration und der Stickstoffdeposition (Depositionsbericht 2017)", Kapitel 1-4, Landesanstalt für Umwelt Baden-Württemberg, Fachdokumentendienst Umweltbeobachtung, ID U46-S7-J16, Ministerium für Umwelt, Klima und Energiewirtschaft Baden-Württemberg [Ed.], Karlsruhe, Germany, http://www.fachdokumente.lubw.baden-wuerttemberg.de/servlet/is/121207/U46-S7-J16.pdf?command=downloadContent&filename=U46-S7-J16.pdf&FIS=91063.

HELCOM (2013), *Summary Report on the Development of Revised Maximum Allowable Inputs (MAI) and Updated Country Allocated Reduction Targets (CART) of the Baltic Sea Action Plan*, Helsinki Commission, Helsinki.

Karl, T. et al. (2017), "Urban Eddy Covariance Measurements Reveal Significant Missing NO_x Emissions in Central Europe", *Scientific Reports*, 7(2536), doi.org/10.1038/s41598-017-02699-9.

Linker, L.C. et al. (2016), "Integrating Air and Water Environmental Management in the Chesapeake Bay Program", presentation to the Joint OECD/TFRN Workshop on The Nitrogen Cascade and Policy – Towards Integrated Solutions, OECD, Paris, 9-10 May 2016.

LUBW (2018), "StickstoffBW", Landesanstalt für Umwelt Baden-Württemberg [Ed.], https://www.lubw.baden-wuerttemberg.de/medienuebergreifende-umweltbeobachtung/stickstoffbw.

LUBW (2016), "Beurteilung der Stickstoffdeposition in Baden-Württemberg -- Kurzmitteilung 1/2016 für eine zwischen Bund und Ländern abgestimmte Stickstoffstrategie", Landesanstalt für Umwelt Baden-Württemberg, Fachdokumentendienst Umweltbeobachtung, ID U10-S7-J16, Ministerium für Umwelt, Klima und Energiewirtschaft Baden-Württemberg and Ministerium für Verkehr und Infrastruktur Baden-Württemberg [Ed.], Karlsruhe, Germany, www.fachdokumente.lubw.baden-wuerttemberg.de/servlet/is/116484/U10-S7-J16.pdf?command=downloadContent&filename=U10-S7-J16.pdf.

Ministry of Economic Affairs (2015), Programma Aanpak Stikstof website, pas.natura2000.nl/ (accessed 27 January 2017).

OECD (2017), *Green Growth Indicators 2017*, OECD Publishing, Paris, doi.org/10.1787/9789264268586-en.

OECD (2015), *Environment at a Glance 2015: OECD Indicators*, OECD Publishing, Paris, doi.org/10.1787/9789264235199-en.

OECD (2012), *OECD Environmental Outlook to 2050: The Consequences of Inaction*, OECD Publishing, Paris, doi.org/10.1787/9789264122246-en.

OECD (2008), *OECD Environmental Outlook to 2030*, OECD Publishing, Paris, doi.org/10.1787/9789264040519-en.

Pinho, P. et al. (2016), "Mapping Portuguese Natura 2000 Sites in Risk of Biodiversity Change Caused by Nitrogen Pollution", contribution to the joint OECD/TFRN Nitrogen Workshop, Paris, 9 10 May 2016, Centre for Ecology, Evolution and Environmental Changes and CERENA, Lisbon, Portugal.

PRIMEQUAL (2015), *Agriculture et Pollution de l'Air : Impacts, Contributions, Perspectives, État de l'Art des Connaissances)*, plaquette résumant un séminaire national organisé le 2 juillet 2014 réunissant des experts du Ministère de l'Écologie, du Développement durable et de l'Énergie, de l'Agence de l'Environnement et de la Maîtrise de l'Énergie et de l'Institut national de la recherche agronomique, Programme de Recherche Inter-organisme pour une Meilleure Qualité de l'Air.

Rabalais, N.N. et al. (2010), Dynamics and Distribution of Natural and Human-caused Hypoxia", *Biogeosciences*, 7.

Roy, R. and N. Braathen (2017), "The Rising Cost of Ambient Air Pollution thus far in the 21st Century: Results from the BRIICS and the OECD Countries", *OECD Environment Working Papers*, N° 124, OECD Publishing, Paris, doi.org/10.1787/d1b2b844-en.

Salomon, M. et al. (2016), "Towards an Integrated Nitrogen Strategy for Germany", *Environmental Science & Policy*, 55 (2016).

SRU (2015), *Nitrogen: Strategies for Resolving an Urgent Environmental Problem*, German Advisory Council on the Environment, Berlin.

Sturdivant, S.K. et al. (2012), "Bioturbation in a Declining Oxygen Environment, *in situ* Observations from Wormcam", *PLoS ONE*, 7(4).

Thunis, P. et al. (2017), *Urban PM$_{2.5}$ Atlas - Air Quality in European Cities*, European Commission, Joint Research Centre, JRC108595, EUR 28804 EN, Publications Office of the European Union, Luxembourg.

Turner, R.E. et al. (1998), "Fluctuating Silicate: Nitrate Ratios and Coastal Plankton Food Webs", *Proc Natl Acad Sci U S A.*, 95(22).

UNEP (2012), *Green Economy in a Blue World*, UNEP, FAO, IMO, UNDP, IUCN, WorldFish Center, GRIDArendal, undp.org/content/dam/undp/library/EnvironmentandEnergy/WaterandOceanGovernance/Green_Economy_Blue_Full.pdf.

USEPA-SAB (2011), *Reactive Nitrogen in the United States: An Analysis of Inputs, Flows, Consequences and Management Options*, U.S. Environmental Protection Agency's Science Advisory Board, EPA-SAB-11-013, USEPA, Washington D.C., yosemite.epa.gov/sab/sabproduct.nsf/WebBOARD/INCFullReport/$File/Final%20INC%20Report_8_19_11(without%20signatures).pdf.

Vieno, M. et al. (2016), "The UK Particulate Matter Air Pollution Episode of March-April 2014: More than Saharan Dust", *Environmental Research Letters*, 11(4).

Chapter 4. The unintended consequences on the nitrogen cycle of conservation practises in agriculture

This chapter warns against the possible unintended effects of nitrogen pollution management measures by providing a case study on agriculture. The chapter reviews the various practices implemented in the United States to manage agricultural nitrogen pollution. The possible unintended effects of each measure are detailed and general lessons are learned.

Regardless of the policy approach (risk, precautionary), it is crucial to consider the reality of the nitrogen cascade to guide nitrogen policy making (see Chapter 1. for a description of the cascade). The design of best nitrogen management practices, and ultimately of instruments to incentivise them, must evaluate their unintended effects on other forms of nitrogen due to the nitrogen cascade. This principle applies to all nitrogen sources such as agriculture, energy, transportation, industry and wastewater treatment.

This Chapter focuses on agricultural practices and argues that they can have unintended consequences if the nitrogen cascade is not taken into account. Many transformations of nitrogen in various forms contribute to its movement between terrestrial, atmospheric, and aquatic ecosystems. Because of this lability, the intended beneficial effects often become unintended detrimental effects for adjacent ecosystems, or even within the ecosystem to which nitrogen is applied (Follett et al., 2010).

4.1 Managing nitrogen for agriculture and the environment

Increased use of commercial nitrogen fertiliser in the United States fuelled an increase in yields, but also posed increased risks to environmental quality. The Natural Resource Conservation Service (NRCS) of the U.S. Department of Agriculture (USDA) evaluated the use of conservation practices on U.S. cropland in major river basins using survey data collected over 2003-06 (USDA-NRCS, 2011a, 2012b,c, 2013a,b). The state of nitrogen management was based on a set of criteria defining appropriate application rate, method of application, and timing of application. Meeting all three of these three criteria would provide adequate nutrients to crops but reduce excess applications that are most at risk for leaving fields and entering air and water.

NRCS found that, on the whole, a great deal of improvement was needed (Table 4.1). The low percentages of good nutrient management practices in the Upper Mississippi, Lower Mississippi and Ohio-Tennessee are noteworthy because this is where a majority of corn is grown. Corn is the largest user of nitrogen, both on a per-acre[1] and total use basis. This region is also a major source of nitrogen entering the Gulf of Mexico via the Mississippi River, which is the primary cause of the large zone of hypoxic water found there (Alexander et al., 2008).

Table 4.1. Percentage of cropland meeting nitrogen rate, method, and timing criteria consistent with "good" management, Mississippi River Basin, 2003-06

River Basin	% cropland
Ohio-Tennessee	17
Upper Mississippi	14
Lower Mississippi	14
Missouri	35
Arkansas-White-Red	33

Source: USDA-NRCS Conservation Effects Assessment Project.

The Clean Air Act and Clean Water Act regulate the environmental impacts of nitrogen in the air and water. USDA conservation programmes promote general improvements in nutrient management through voluntary programmes. Decisions on which conservation measures to adopt are made by individual farmers. In protecting a particular environmental medium from agricultural pollution, specific sets of management practices are promoted, either through regulation (rare in the United States) or financial assistance and education (common). The choice of practices to support is often made based on their expected effectiveness for preventing pollutant losses to a particular media. The focus on one pathway can result in unintended consequences that are detrimental to other media.

In the following section the movement of nitrogen is traced for crop production and for confined animal operations. Both are important sources of nitrogen to the environment, with different management approaches.

4.2 Nitrogen pathways in crop production

Emissions of nitrogen to water and to the atmosphere are not independent events, but are linked by the biological and chemical processes that produce the various nitrogen forms. Crop production is characterised by stochastic weather and soil conditions that affect crop yields and nitrogen loss. From a nutrient standpoint, crop production is a "leaky" system. It is impossible to ensure that every bit of nutrient input to cropland via direct application of commercial fertiliser or animal waste, or fixed by legumes, is taken up by the planted crop. Nitrogen applied to cropland can be "lost" in a variety of ways:

- *Soil erosion* - Nitrogen can be lost from the soil surface when attached to soil particles that are carried off the field by wind or water. Although wind and water erosion can be observed across all regions, wind erosion is more prevalent in dry regions and water erosion in humid regions. Overall, little nitrogen is lost through erosion when basic conservation practices are in place (Iowa Soybean Association, 2008).

- *Runoff* – Surface runoff of dissolved nitrogen, generally in the form of nitrate (NO_3^-), is only a concern when fertiliser and or manure are applied on the surface and rain moves the nitrogen before it enters the soil (Legg and Meisinger, 1982; Iowa Soybean Association, 2008).

- *Leaching* - Leaching occurs when there is sufficient rain and/or irrigation to move easily dissolvable NO_3^- through the soil profile (Randall et al., 2008). NO_3^- eventually ends up in underground aquifers or in surface water via tile drains and groundwater flow. Tile drains may be a chief passageway by which nitrogen moves from crop soils to surface water in regions with high water tables (Turner and Rabalais, 2003; Randall et al., 2008; Randall et al., 2010; Petrolia and Gowda, 2006).

- *Ammonia (NH_3) volatilisation* - Significant amounts of nitrogen can be lost to the atmosphere as NH_3 if animal manure or urea is surface applied and not immediately incorporated into the soil (Hutchinson et al., 1982; Fox et al., 1996; Freney et al., 1981; Sharpe and Harper, 1995; Peoples et al., 1995). Additionally, warm weather conditions can accelerate the conversion of manure and other susceptible inorganic nitrogen fertilisers to NH_3.

- *Denitrification* - When oxygen levels in the soil are low, some microorganisms, known as denitrifiers, will convert NO_3^- to dinitrogen and nitrous oxide (N_2O) (Mosier and Klemedtsson, 1994). Denitrification usually occurs when NO_3^- is present in the soil, soil moisture is high or there is standing water, and soils are warm. The ratio of dinitrogen to N_2O is governed by the amount of oxygen available to the denitrifying organisms. The higher the level of oxygen, the greater the amount of N_2O produced.

To maintain economically viable farming operations, farmers manage temporal variability in weather and soil nitrogen by applying more nitrogen than plants need to protect against downside risk (i.e. use an "insurance" nitrogen application rate) (Sheriff, 2005; Babcock, 1992; Babcock and Blackmer, 1992; Rajsic and Weersink, 2008). This ensures that nitrogen needed by the crop is available. Additionally, farmers may take a "safety net" approach to maximise economic returns by setting an optimistic yield goal for a given field based on an optimum weather year to ensure that the amount of nitrogen needed for maximum yields is available (Schepers et al., 1986; Bock and Hergert, 1991). Thus, during years in which weather is not optimal for maximising yields, nitrogen will be over-applied from an agronomic standpoint. Almost by definition, optimal conditions are infrequent, so farmers following this approach over-fertilise crops in most years. The decision to apply "extra" nitrogen is economically justified if the cost of over-applications is low compared to the cost of under-application (Rajsic and Weersink, 2008).

4.3 Nitrogen pathways in animal production

Nitrogen cycling in animal production adds the dimension of manure production, storage and management. Nitrogen enters the system in animal feed. Some of the nitrogen is retained in the animal products (meat, milk, eggs), but as much as 95% is excreted in urine and manure, much of which is applied to cropland as fertiliser (Follett and Hatfield, 2001).

Manure can collect in or under the production house for a few hours or several months, depending on the collection system. Production houses are ventilated to expel gases that are emitted, including NH_3. The manure is eventually removed from the house to a storage structure (lagoon, tank, pit, or slab) and stored anywhere from a few days to many months. Losses of nitrogen to air and water can occur during this time, depending on the system and the extent of contact with rain and wind. The stored manure is eventually transported to fields where it is applied. Losses to air and water from the field vary, depending on application method and rate. Nitrogen in the field helps produce crops, which may in turn be fed to animals, thus completing the cycle. Nitrogen lost to the air eventually returns to earth, where it can be a source of plant nutrients, or be lost, as decribed above.

The form nitrogen takes in its journey from animal to field depends on a host of factors, including storage technology, manure moisture content, temperature, air flow, pH, and the presence of micro-organisms. Reducing nitrogen movement along one path by changing its form will increase nitrogen movement along a different path (NRC, 2003). For example, reducing NH_3 losses from a field to the atmosphere by injecting waste directly into the soil increases the amount of nitrogen at risk of moving to water resources as NO_3^- (Oenema, 2001; Abt Associates, 2000). Ignoring the interactions of the nitrogen cycle in developing manure management policies could lead to unintended and adverse effects on environmental quality.

4.4 Conservation practices and the nitrogen cycle

Standard conservation practices have been evaluated for their impacts on nitrogen losses along different pathways. Changes to crop rotations as a means of reducing nitrogen losses are not considered, other than the addition of a cover crop.[2] The following practices influence nitrogen losses to the environment. Their effectiveness varies tremendously across the setting (crop, soil, climate, management skill) in which they are applied. The discussion is therefore limited to general impacts. It is recalled that practices designed to achieve specific environmental policy objectives are often used in combination (Box 4.1).

> **Box 4.1. Conservation practices are often used in combination**
>
> Nitrogen pollution problems cannot usually be solved with one management practice because single practices do not typically provide the full range and extent of control needed at a site. Multiple practices are combined to build management practice systems, which are generally more effective in controlling the pollutant since they can be used at two or more points in the pollutant delivery process. On the other hand, a set of practices does not constitute an effective management practice system unless the practices are selected and designed to function together to achieve specific environmental policy goals. For example, if water quality is the environmental policy goal, nutrient management, conservation tillage, field borders, and riparian buffers can be associated to control nitrogen losses. Conservation tillage can help reduce overland transport of nitrogen by reducing erosion and runoff, and nutrient management will minimize subsurface losses due to the resulting increased infiltration. Filter strips can be used to decrease nitrogen transport by increasing infiltration, and through uptake of available nitrogen by the field border crop. Nitrogen not controlled by nutrient management, conservation tillage, and filter strips can be intercepted and remediated through denitrification in riparian buffers.
>
> Each practice in a management practice system must be selected, designed, implemented, and maintained in accordance with site-specific considerations to ensure that the practices function together to achieve the overall management goals. If, for example, nutrient management, conservation tillage, filter strips, and riparian buffers are used to address a water quality problem, then planting and nutrient applications need to be conducted in a manner consistent with conservation tillage goals and practices (e.g. injecting rather than broadcasting and incorporating fertiliser). In addition, runoff from the fields must be conveyed evenly to the filter strips which, in turn, must be capable of delivering the runoff to the riparian buffers in accordance with design standards and specifications.
>
> Source: USEPA (2003).

4.4.1 Nutrient management

Nutrient management is defined by USDA's NRCS as managing the amount, placement, form, and timing of the application of plant nutrients to the soil (USDA-NRCS, 2006). The focus is on overall nutrient use efficiency (NUE),[3] and not on any particular vector or pathway of nutrient loss to the environment. But decisions related to the rate, timing, and method of application have implications for the form and pathway of nitrogen movement in and out of the soil.

Nitrogen can be applied to cropland in a variety of forms. Some of the more widely used nitrogen fertiliser forms include anhydrous NH_3 (gas), urea (solid), Urea Ammonium Nitrate (liquid),[4] and manure (solid). These forms vary in how quickly they can be transformed into NO_3^-, which is what crops actually use. The closer in time the fertiliser is applied to when the crop needs it, the faster it needs to be transformed into NO_3^-. A mismatch of fertiliser form with appropriate timing can lead to large environmental losses and poor yields.

NUE can be increased by placing fertilisers in the soil rather than broadcasting them on the surface. Such placement reduces the risks of losses to the atmosphere or to surface runoff. Subsurface placement of nitrogen also reduces losses stemming from NH_3 volatilisation. However, some studies have found that incorporation into the soil increases N_2O emissions (Flessa and Beese, 2000; Wulf et al., 2002; Drury, 2006). Injection or incorporation could also increase NO_3^- leaching, especially where soils are coarse textured (Abt Associates, 2000).

NUE can be increased by improving the synchronisation between crop nitrogen demand and the supply of nitrogen throughout the growing season (Doerge et al., 1991; Cassman et al., 2002; Meisinger and Delgado, 2002). This implies maintaining low levels of inorganic nitrogen in the soil when there is little plant growth and sufficient inorganic nitrogen fertiliser during periods of rapid plant growth. For example, the corn plant's need for nitrogen is not very high until about 4 weeks after it emerges from the ground, which typically falls in June through July in the major corn producing regions of the United States. However, farmers often apply nitrogen in the fall for a spring-planted crop, due to lower fertiliser prices and time constraints in the spring.

Cold soils and the use of nitrogen inhibitors can slow the conversion of nitrogen fertiliser to NO_3^-, but losses to leaching can be quite substantial. Switching to spring applications can reduce overall application rates and NO_3^- loss to water, but may increase N_2O emissions as applications are shifted to generally warmer, wetter conditions (Delgado et al., 1996; Rochette et al., 2004; and Hernandez-Ramirez et al., 2009). Applying nitrogen during the growing season also introduces some risk, as weather may prevent an application when plants need it, negatively affecting yields. Whilst nitrification inhibitors are effective at decreasing direct N_2O emission and NO_3^- leaching, recent studies suggest that they may increase NH_3 volatilisation and, subsequently, indirect N_2O emission (Lam et al., 2017).

Precision agriculture technologies are playing an increasing role in farm production, creating more opportunities for increasing NUE. Precision technologies – such as tractor guidance systems using a global positioning system (GPS), GPS soil and yield mapping, and variable-rate input applications (VRT) – help farms gather information on changing field conditions to adjust fertilisation practices (Schimmelpfennig, 2016). The first wave of precision agriculture was GPS guidance for tractors; introduced in the early 1990s it is now widespread (Schmaltz, 2017). GPS reduces driving errors and any overlap of fertiliser spreading. Uptake of the second technological wave, VRT, is currently estimated at 15% in North America and is expected to continue to grow rapidly (ibid). The third wave may well be that of "Big Data" and the fourth, in the longer term, that of robotics. Big Data are massive volumes of data with a wide variety that can be captured (e.g. by drones), analysed and used for decision-making (Wolfert et al., 2017).

4.4.2 Tillage

Conservation tillage is a popular practice that has several attractive features, including reduced fuel use, more available soil water, and reduced soil erosion. Reducing soil erosion keeps soil particles and attached nutrients out of surface water. In some studies, no-till systems have been shown to reduce NO_3^- leaching over conventional tillage, as well as proper crop rotation, especially those including a nitrogen-fixing crop (Meek et al., 1995). However, another study has shown that conservation tillage increases the infiltration rate of soils (Baker, 1993). The potential for NO_3^- leaching may increase due to greater availability of water and increased soil porosity (i.e. large pore spaces) (USEPA, 2003). Soil macroporosity and the proportion of rainfall moving through preferential flow paths often increase with the adoption of conservation tillage, potentially increasing the transmission of NO_3^- and other chemicals available in the upper soil to subsoils and shallow groundwater (Shipitalo et al., 2000). Also, reduced tillage creates an environment more favourable to ammonification and denitrification, which could lead to increased emission of nitrogen forms to the atmosphere. However, this depends on soils and soil moisture (MacKenize, Fan, and Cadrin 1997).

4.4.3 Cover crops

Cover crops are planted after the principal crop has been harvested to provide soil surface cover and reduce soil erosion, and to prevent nitrogen from leaching or running off to surface water (Dabney et al., 2010). Cover crops do this by taking up water and NO_3^-, thus reducing the volume of NO_3^- percolating through the soil. For example a 6-year (1999-2005) study on corn and soybeans in Canada showed that planting a winter wheat cover crop could reduce NO_3^- losses relative to no cover crop while increasing yields (Drury et al., 2014). Cover crops have the added benefit of building up soil carbon. However, cover crops are not as effective at reducing atmospheric losses of nitrogen, as these occur primarily during or shortly after fertiliser applications.

4.4.4 Filter strips

Vegetative filter strips remove sediment, organic matter, and other pollutants from surface runoff, thus keeping them out of streams. The impact on dissolved nitrogen is less clear. Denitrification may be enhanced by the filter. Also, NO_3^- leaching may increase because surface water movement is slowed. Fares et al., 2010 review the literature on vegetative filters and the nitrogen cycle.

4.4.5 Restored wetlands

Restored wetlands have been shown to be effective at reducing the movement of nitrogen from cropland to surface water (Jansson et al., 1994; Hey et al., 2005; Mitsch and Day, 2006). Wetland vegetation takes up nitrogen, and wet soils enhance denitrification. Most of nitrogen removal occurs through denitrification (Crumpton et al., 2008). There is some support for the belief that denitrification produces more of the inert dinitrogen and little N_2O (USEPA, 2010; Hernandez and Mitsch, 2006; Crumpton et al., 2008).

4.4.6 Field drainage

Tile drainage is used to reduce runoff and increase soil drainage. It lowers the water table of fields that would otherwise be too wet to crop intensively. Drained soils tend to be highly productive. About 26 % of cropland receiving nitrogen is tiled, most of this in corn production (Ribaudo et al., 2011). However, about 71 % of tiled cropland acres do not meet the three nitrogen management criteria.

Tiles provide a rapid conduit for soluble NO_3^-, effectively bypassing any attenuation that may occur in the soil or in vegetative buffers. As such, tile drains can have the undesirable effect of concentrating and delivering nitrogen directly to streams (Hirschi et al., 1997). NO_3^- losses from fields with tile drainage are a major source of water quality degradation in areas where tiled fields are present (Dinnes et al., 2002; David et al., 2010, Randall et al., 2010).[5]

In order to reduce the nitrogen pollution caused by tile drains, other management practices, such as nutrient management for source reduction and biofilters (or bioreactors) that are attached to the outflow of the tile drains for interception, may be needed (USEPA, 2003). However, bioreactors also enhance denitrification.[6]

On the other hand, practices which control runoff may contribute to reduced in-stream flows (USEPA, 2003). One practice that is being used to control NO_3^- losses is drainage water management (DWM). DWM is the process of managing the timing and the amount of water discharged from agricultural drainage systems. During the growing season water levels are lowered (drainage increased) to allow optimal crop growth. During fallow periods the water level is allowed to rise, which reduces the volume of drainage water leaving a field and enhances denitrification within the soil profile (Randall et al., 2010).

4.4.7 Chemical additions to manure

Chemicals such as alum can be added to manure during its collection in order to bind odorous compounds and to reduce NH_3 emissions (Moore et al., 2000). By doing so the nitrogen content of manure that is spread on fields is increased. This could pose an increased risk for NH_3 and NO_3^- losses from the field, unless appropriate field management practices are employed.

4.4.8 Tank covers

Covering manure storage tanks can greatly reduce the discharge of NH_3 to the atmosphere, primarily by altering pH and preventing the formation of NH_3 (Jacobson et al., 1999). While reducing NH_3 emissions, covers also increase the nitrogen content of effluent that is eventually spread on fields, increasing the potential for both NH_3 emissions and loss of NO_3^- through leaching unless appropriate field management practices are employed.

4.4.9 Slurry lagoon covers

Plastic covers that float on the lagoon surface or that are tented over lagoons can greatly decrease the emission of gaseous nitrogen forms (Jacobson et al., 1999; Arogo et al., 2002). Some systems (anaerobic digesters) also capture methane and use it as a biofuel to generate electricity. Covering a lagoon increases the nitrogen content of the effluent that is eventually sprayed on fields. While NH_3 emissions from fields sprayed with lagoon effluent might increase, the net effect is a reduction in NH_3 emissions from both lagoon storage and field applications. However, the risk of NO_3^- loss to water increases.

4.4.10 Manure incorporation and injection

About 10% of crop acres treated with nitrogen received animal manure in 2006 (Ribaudo et al., 2011). Whereas 62 % of the cropland acres receiving only commercial fertiliser were not meeting the criteria for good nitrogen management, 86 % of cropland acres receiving only manure were not meeting the criteria, and 96 % of the cropland receiving both commercial fertiliser and manure were not meeting the criteria. Research has indicated that farms with confined animals tend to over-apply nutrients to crops, primarily because of the large amount of manure produced relative to available cropland on the farm (Ribaudo et al., 2003; Gollehon et al., 2001).

Rapidly incorporating manure into the soil, either by plowing or disking solids into the soil after spreading, or by injecting liquids and slurries directly into the soil, reduces NH_3 emissions (Abt, 2000; Arogo et al., 2002). But this also increases the nitrogen available for crops in the soil, and thus the risk of NO_3^- leaching and runoff to water bodies. Therefore, manure incorporation and injection needs to be accompanied by appropriate soil testing and other field management practices.

4.5 Changing nutrient management on cropland may result in environmental trade-offs

A study by USDA's Economic Research Service (ERS) (Ribaudo et al., 2011) assessed how the adoption of different management practices affected the movement of nitrogen along different pathways by using the Nitrogen Loss and Environmental Assessment Package (NLEAP) model with Geographic Information System (GIS) capabilities (Shaffer et al., 2010; Delgado et al., 2010). Of particular interest in this research was the extent to which trade-offs in environmental outcomes might occur as overall NUE is improved.

Because NLEAP is a field-level model, eight different soils in four States (Arkansas, Ohio, Pennsylvania, and Virginia) were selected to assess changes in nitrogen emissions to the environment from management changes in non-irrigated corn production.[7] Four of the soils were type A or B soils (well drained), and four were type D soils (relatively poorly drained). For each soil, two rotations (corn-corn and corn-soybeans), two tillage practices (conventional and no-till), and two sources of nitrogen (inorganic fertiliser and inorganic fertiliser + animal manure) were examined. The slopes for these soils were 0 to 6%, with low erosion potential.

For each soil/rotation/tillage/nitrogen-source combination, eight different scenarios were modelled with NLEAP. Starting from a baseline where at least one of the criteria for good nitrogen management is not being met (right rate, right timing, and right application method), changes in management are made so that all criteria are met. Changes in only timing and/or method of application assumed that application rate did not change.

All the scenarios show the expected changes in total nitrogen losses as the criteria were met, with reductions indicating improvements in NUE. The NLEAP results were consistent with the expectation that nitrogen emissions are minimised when all three criteria are met. Since nitrogen cascades through different forms and ecosystems, the long-term environmental benefits of reducing total nitrogen are clear. However, some of the trade-offs between different forms of nitrogen could pose environmental problems. In our example, adopting injection/incorporation always increased NO_3^- leaching, sometimes substantially (more than doubling leaching in some cases). Similarly, shifting applications from fall to spring (without changing application rate) reduces NO_3^- losses and total nitrogen losses but increases N_2O emissions as applications are shifted to generally warmer, wetter conditions, which is consistent with the findings of Delgado et al., 1996, Rochette et al., 2004, and Hernandez-Ramirez et al., 2009. Due to the potential for increasing greenhouse gas (GHG) emissions, this outcome would have to be carefully considered when making recommendations to improve NUE.

Injection/incorporation and eliminating fall applications again produced mixed results. NH_3 emissions are always reduced. Leaching is generally reduced, but in some cases where manure is used, it may increase. N_2O emissions almost always increase, from 5 to 50%, depending on the situation. Only reducing the application rate guarantees that losses of all three forms of nitrogen are reduced (Mosier et al., 2002; Meisinger and Delgado, 2002). This suggests that in areas where leaching to drinking water sources is possible, improvements in NUE could focus on application rate reductions or improvements in timing.

4.5.1 NRCS Conservation Effects Assessment Project

These trade-offs can also be observed at a larger scale. NRCS evaluated the effects of conservation systems (tillage, nutrient management, erosion controls) on nitrogen loss from cropland in 12 large watersheds in the United States as part of the Conservation Effects Assessment Project (CEAP) (USDA-NRCS, 2011a,b; 2012 a,b,c; 2013 a,b,c; 2014 a,b,c,d; 2015). The emphasis on nutrient management by USDA has been to reduce losses to water. Through a combination of field surveys and modelling, the fraction of nitrogen applied to cropland that is taken up by crops (a measure of NUE), as well as losses through different pathways were estimated for cropping practices observed during the years 2003-06.[8] In general, the percentage of nitrogen application that is taken up by crops ranged from 60% in the South Atlantic Gulf to 76% in the Souris-Red-Rainy (Table 4.2).

Table 4.2. Changes in nitrogen loss to the atmosphere with the adoption of nutrient management practices, by major watershed

Watershed	Crop uptake of nitrogen applied (%)	Change in per-acre nitrogen loss through volatilisation, with vs. without observed conservation measures (%)	Change in per-acre nitrogen loss through denitrification, with vs. without observed conservation measures (%)
Delaware River	72	1	16
Chesapeake Bay, 2006	62	-4.2	1.2
Chesapeake Bay, 2006-2011	66	3.2	1.9
South Atlantic Gulf	60	23	32
Great Lakes	72	7	-6
Ohio-Tennessee	73	21	0
Upper Mississippi	72	3	5
Lower Mississippi	64	10	45
Texas Gulf	57	-11	-9
Missouri	75	-11	-4
Arkansas-White-Red	72	-31	-34
Souris-Red-Rainy	76	-28	-28
Pacific Northwest	72	-37	-26

Source: USDA-NRCS (2011 a, b; 2012 a, b, c; 2013 a, b, c; 2014 a, b, c, d; 2015).

NRCS estimated through modelling how observed management practices affect nitrogen loss assuming an alternative state of no conservation practices. The results are summarised in Table 4.2. In some major agricultural watersheds, such as Upper and Lower Mississippi and Ohio-Tennessee, the observed mix of conservation practices were estimated to have increased the amounts of nitrogen lost to the atmosphere due to volatilisation and denitrification (3 % and 5 %, respectively) even though total nitrogen losses were reduced (NUE increased). This is generally because nitrogen fertiliser remained on the field longer, where it is exposed to wind and weather conditions that promote volatilisation and denitrification. On the other hand, nitrogen losses to all pathways were estimated to have been reduced by the observed management systems.

The results for the Chesapeake are particularly instructive. A field survey was conducted both in 2006 and 2011, spanning a time period when there was great emphasis on reducing nutrient loads to the Chesapeake Bay. Between 2006 and 2011, NUE was estimated to have increased from 62% to 66%, yet nitrogen losses to volatilisation (NH_3) and denitrification (primarily dinitrogen but also N_2O) were estimated to have increased 3.2% and 1.9%, respectively. The management practices chosen to reduce nitrogen losses to water (nutrient management, cover crops, and erosion controls) appeared to have resulted in increased losses to the atmosphere.

An analysis of the pathways of nitrogen in air and water would inform trade-off management between air pollution and water pollution. Following the Chesapeake Bay example, about half of the NH_3 emission from the Chesapeake watershed is not returning to the watershed because of atmospheric transport (Linker et al., 2013). It is estimated that 90% of the load deposited on the watershed is attenuated in the forests, fields, and other land uses, and about half of the remaining 10% is lost in transport in rivers to the tidal waters of the Bay (ibid). In other words, of 100 kg of NH_3 lost from volatilisation from the Bay cropland only an estimated 2 kg make it to the eutrophic waters of the tidal Chesapeake. Further, NH_3 emissions are generally not controlled by any Clean Air Act regulations.

4.6 Water-air trade-offs in manure management

Clean Water Act regulations require concentrated animal feeding operations (CAFOs) to meet a nutrient standard for land application in order to keep NO_3^- out of ground and surface water. This generally means more land is required for spreading manure than has been used in the past, an expensive proposition for many large farms (Ribaudo et al., 2003). One rational management response under a nitrogen standard is to encourage volatilisation of manure nitrogen (e.g. use of uncovered lagoons, surface application to fields) to reduce the nitrogen content in waste, thus allowing higher application rates on cropland and reducing the amount of land needed for spreading (Sweeten et al., 2000). But, such a strategy would increase atmospheric emissions of NH_3 and worsen air quality. Zilberman et al., 2001 cite the multi-path nature of animal waste as one reason why policies are inadequate. A policy focused on nitrogen applications to land also allows the build-up of other potential pollutants in the soil, such as phosphorus,[9] and ignores problems such as odour and dust.

An ERS study evaluated farm-level responses by hog producers of environmental restrictions on nitrogen emissions to water, nitrogen emissions to air, and nitrogen emissions to both water and air. This analysis demonstrated the potential unintended consequences of a policy that ignores the nitrogen cycle (Aillery et al., 2005).

The water protection policy considered by the ERS study was a nitrogen application standard that limited applications to a maximum of 50% more than crop uptake. The largest cost faced by a confined animal operation characterised by excess nutrient applications when confronted with a nitrogen application limit is the cost of hauling manure to a larger land base (Ribaudo et al., 2003). Hog operations met the restriction by spreading on more land, and for operations storing manure in tanks, decreasing slurry injection by 11.8% (thus enhancing NH_3 emissions and reducing the nitrogen content of the slurry). Overall, NH_3 emissions increased by 3.4%, mainly because more land is receiving manure and because of the switch by some producers from injection to surface application.

The ERS-modeled air policy was a limit on NH_3 emissions from pit operations at 10% above the minimum obtained if all manure is injected. For lagoon operations, NH_3 emissions were constrained to 20% above what is obtainable if lagoons were covered. The constraint on NH_3 emissions is in the form of a percentage reduction in net nitrogen emissions per pig. The NH_3 limits induced pit operations to switch from surface application to injection on some land, and induced some lagoon operations to cover their lagoons. The NH_3 limits resulted in a 38% decline in NH_3 emissions from manure storage facilities (the largest source of NH_3 emissions), and a 57% increase in NH_3 emissions from fields, for a net decline in emissions of 29%. The increase in emissions from fields results because more lagoons are covered, which raises the nutrient content of the lagoon liquid applied to fields, resulting in greater nitrogen volatilisation. The NH_3 standard resulted in a 79% increase in excess nitrogen applied to soil[10] (greatly increasing threats to water) – revealing an important trade-off between water and air quality.

4.7 Nitrous oxide (N_2O) management practices

Recent studies have examined how N_2O management practices could be a cost-effective lever for mitigating agricultural GHG emissions. For example, Pellerin et al., 2013 identified practices that could reduce N_2O emissions without entailing major changes in production systems or a significant reduction in yield (less than 10%) and estimated their mitigation potential and their cost over 2010-30. The study shows that eight practices are cost-effective in reducing N_2O emissions (Table 4.3). Since the final effect of these practices is a reduction of the amount of nitrogen applied on fields the risk of nitrogen pollution swapping is minimised.

Table 4.3. Practices to reduce nitrous oxide (N₂O) emissions

Practices	Sub-practices
Reduce the application of mineral nitrogen fertilisers to reduce the associated N₂O emissions	
Reduce the use of synthetic mineral fertilisers through their more effective use and making greater use of organic resources	Adjust nitrogen fertiliser application rates to more realistic yield targets* Make better use of organic fertiliser* Adjust application dates to crop requirements* Add a nitrification inhibitor Incorporate fertiliser to reduce losses*
Increase the use of legumes to reduce the use of synthetic nitrogen fertilisers	Introduce more grain legumes in arable crop rotations Increase legumes in temporary grassland*
Store carbon in soil and biomass	
Introduce more cover crops, intercropping and green cover strips in cropping systems	Introduce more cover crops Introduce more vineyard/orchard cover cropping
Optimise grassland management to promote carbon storage	Make the most intensive permanent and temporary grassland less intensive by more effectively adjusting nitrogen fertiliser application*
Modify the animal diet to reduce enteric methane emissions and N₂O emissions related to manure	
Reduce the amount of protein in the livestock diet to limit the quantity of nitrogen excreted in manure and the associated N₂O emissions	Adjust nitrogen content in the diet of dairy cows* Adjust nitrogen content in the diet of pigs*

Note: * Cost-effective measure (i.e. with negative annual cost per tonne of CO₂e avoided).
Source: Pellerin et al. (2013).

4.8 Summary, conclusions and areas for further analysis

Controlling nitrogen emissions from field applications without considering the nitrogen cycle can lead to unintended consequences. For example, injecting/incorporating nitrogen inorganic and organic fertiliser into the soil reduces atmospheric losses, but can increase NO_3^- leaching. Fertilising only during the growing season can reduce NO_3^- loss to water but may increase N_2O emissions. Reducing applied nitrogen reduces losses via all pathways.

Tillage and vegetative filters can also lead to unintended consequences. Conservation tillage reduces runoff of nutrients and sediment but may increase NO_3^- leaching and atmospheric losses. Vegetative filter strips remove sediment, organic matter, and other pollutants from surface runoff but may increase NO_3^- leaching and atmospheric losses.

Field drainage is a major source of NO_3^- losses to water. Drainage water management reduces the volume of drainage water and pollutants leaving a field but the practice can enhance denitrification and atmospheric losses of N_2O. Bioreactors reduce NO_3^- concentrations in field drainage by enhancing denitrification but losses of N_2O to the atmosphere could increase.

Manure management decisions affect nitrogen losses. Tank/lagoon covers reduce odour and atmospheric losses of NH_3 but increase the nitrogen content of the material that is eventually applied to fields, increasing the risk of loss during application as well as the cost. Chemical additions to manure such as alum reduce odour and NH_3 emissions but also increase the nitrogen content of material applied to fields.

NRCS CEAP evaluated the effects of conservation systems (tillage, nutrient management, erosion controls) on nitrogen losses in 12 large US river basins. Conservation systems reduced nitrogen losses to water, but in certain circumstances increased nitrogen losses to the atmosphere at the watershed scale.

ERS modelled an NH_3 emission limit on hog operations without restrictions on land applications. The ERS study revealed that reducing NH_3 emissions from manure can increase NO_3^- losses.

A recent study shows that the ultimate effect of cost-effective practices for reducing N_2O emissions is a reduction in the amount of nitrogen applied to the fields. All the risks of pollution by agricultural nitrogen are thus minimised.

In summary, trade-offs between air and water are prevalent in agricultural nitrogen management. Unintended consequences of uncoordinated farming practices can lessen overall environmental gains. Uncoordinated farming practices could impose extra costs on farmers. Reducing nitrogen at the source (fertiliser and manure over-application) could address multiple problems. The amount of manure nutrients that needs to be spread on land could be reduced through feed management and alternative uses for manure.

In conclusion, the multi-pollutant nature of nitrogen flows in the environment suggests there are benefits to coordination of practices, rather than a piecemeal control (Bull and Sutton, 1998; Baker et al., 2001). Both field and regional studies have shown focusing on one particular environmental media, such as water quality, can create incentives for management that result in degradation in a different media (air). Trade-offs might not be an issue if one media can absorb an increase in nitrogen loads. But if increases are undesirable, then farming practices that consider all pathways of nitrogen loss are more efficient than farming practices that do not.

In general, not creating pollution in the first place avoids the problems posed by trying to control different nitrogen pathways. In the case of cropland, reducing nitrogen application rates to achieve a better NUE reduces nitrogen loss along all pathways, avoiding the trade-offs characteristic of most other conservation practices. In the case of nitrogen and confined animal operations, increasing the efficiency of nutrient conversion to animal products can reduce nitrogen in the waste. This would reduce the threats to air and water quality, and make addressing either by managing manure less costly.

The analysis in this Chapter focuses on selected nitrogen management practices. Other management practices could be analysed, such as crop rotation, livestock feed management or irrigation. Assessing the specificity of nitrogen-fixing crops and organic farming could also be of interest. On the latter, for example, there appears to be no scientific consensus that N_2O emissions are lower in soils receiving manure compared to soils receiving inorganic nitrogen fertilisers (Graham et al., 2017).

Notes

[1] One acre equals 0.4 hectare.

[2] Also called « catch crop ».

[3] Defined as the fraction of nitrogen applied to cropland that is taken up by crops.

[4] Urea Ammonium Nitrate (UAN) is a liquid fertiliser consisting of a blend of ammonium, nitrate and urea.

[5] Much of the tile-drained cropland is located in the Mississippi River Basin, which has implications for hypoxia in the Gulf of Mexico.

[6] A bioreactor is a structure at the end of a tile line that is filled with wood chips or other material that promotes denitrification (Christianson et al., 2012).

[7] These four States were selected because they present a wide variation in growing conditions and because the data necessary for running NLEAP were already developed.

[8] For example, in the Upper Mississippi Basin, given the observed mix of management systems, the average amount of total nitrogen lost from a field to the environment was about 39 pounds per acre (USDA-NRCS, 2012c). This is parsed among the nitrogen pathways as follows: 6.9 lbs/acre due to volatilisation; 2.3 lbs/acre due to denitrification; 2.1 lbs/acre due to windborne sediment; 8.8 lbs/acre due to surface runoff; 18.7 lbs/acre due to subsurface flow.

[9] Because the ratio of phosphorus to nitrogen in manure is greater than the ratio needed by plants, meeting a nitrogen application standard allows the continued over application of phosphorus.

[10] Increasingly stringent NH_3 reductions in the US hog sector increase the amount of excess nitrogen applied to fields (Aillery et al., 2005).

References

Abt Associates (2000), *Air Quality Impacts of Livestock Waste*, Report prepared for US Environmental Protection Agency, Office of Policy, Bethesda, MD.

Aillery, M. et al. (2005), *Managing Manure to Improve Air and Water Quality*, Economic Research Report, ERR-9, Economic Research Service, U.S. Department of Agriculture.

Alexander, R.B. et al. (2008), "Differences in Phosphorus and Nitrogen Delivery to the Gulf of Mexico from the Mississippi River Basin", *Environmental Science & Technology*, 42.

Arogo, J. et al. (2002), *Ammonia Emissions from Animal Feeding Operations*, National Centre for Manure and Animal Waste Management, North Carolina State University, Raleigh, NC.

Babcock, B.A. (1992), "The Effects of Uncertainty on Optimal Nitrogen Applications", *Review of Agricultural Economics*, 14(2).

Babcock, B.A. and A.M. Blackmer (1992), "The Value of Reducing Temporal Input Nonuniformities", *Journal of Agricultural and Resource Economics*, 17.

Baker, D.B. (1993), "The Lake Erie Agroecosystem Program: Water Quality Assessments", *Agriculture, Ecosystems and Environment*, 46.

Baker, L.A. et al. (2001), "Multicompartment Ecosystem Mass Balances as a Tool for Understanding and Managing the Biogeochemical Cycles of Human Ecosystems", *TheScientificWorld*, 1(S2).

Bock, B.R. and W. Hergert (1991), "Fertilizer Nitrogen Management", in Follet H. et al. (eds.), *Managing Nitrogen for Groundwater Quality and Farm Profitability*, Soil Science Society of America.

Bull, K.R. and M.A. Sutton (1998), "Critical Loads and the Relevance of Ammonia to an Effects-Based Nitrogen Protocol", *Atmospheric Environment*, 32(3).

Cassman, K.G. et al. (2002), "Agroecosystems, Nitrogen-Use Efficiency, and Nitrogen Management", *Ambio*, 31(2).

Christianson, L.E. et al. (2012), "A Practice-oriented Review of Woodchip Bioreactors for Subsurface Agricultural Drainage", *Applied Engineering in Agriculture*, 28 (6).

Cowling, E. et al. (2002), "Optimizing Nitrogen Management and Energy Production and Environmental Protection", presented at the 2nd International Nitrogen Conference, Bolger Centre, Potomac, MD, 14-18 October 2001, www.initrogen.org/fileadmin/user_upload/Second_N_Conf_Report.pdf.

Crumpton, W. G. et al. (2008), "Potential of Restored and Constructed Wetlands To Reduce Nutrient Export From Agricultural Watershed in the Corn Belt", in Final Report: *Gulf Hypoxia and Local Water Quality Concerns Workshop*, American Society of Agricultural and Biological Engineers.

Dabney, S.M. et al. (2010), "Using Cover Crops and Cropping Systems for Nitrogen Management", in Delgado J.A. and R.F. Follet (eds.), *Advances in Nitrogen Management for Water Quality*, Soil and Water Conservation Society.

Delgado, J.A. et al. (1996), "Effects of N Management on N_2O and CH_4 Fluxes and N Recovery", *Nutrient Cycling in Agroecosystems*, 46.

Delgado, J.A. et al. (2010), "A New GIS Nitrogen Trading Tool Concept for Conservation and Reduction of Reactive Nitrogen Losses to the Environment", *Advances in Agronomy*, 105.

Dinnes, D.L. et al. (2002), "Nitrogen Management Strategies to Reduce Nitrate Leaching in Tile-Drained Midwestern Soils", *Agronomy Journal*, 94.

Doerge, T.A. et al. (1991), *Nitrogen Fertilizer Management in Arizona*, University of Arizona, College of Agriculture.

Drury, C.F. et al. (2006), "Emissions of Nitrous Oxide and Carbon Dioxide: Influence of Tillage Type and Nitrogen Placement Depth", *Soil Science Society of America Journal*, 70(2).

Drury, C.F. et al. (2014), "Reducing Nitrate Loss in Tile Drainage Water with Cover Crops and Water-table Management Systems", *J Environ Qual.*, 43(2), doi.org/10.2134/jeq2012.0495.

Fares, A. et al. (2010), "Use of Buffers to Reduce Sediment and Nitrogen Transport to Surface Water Bodies," in Delgado J.A. and R.F. Follett (eds.), *Advances in Nitrogen Management for Water Quality*, Soil and Water Conservation Society.

Flessa, H. and F. Beese (2000), "Laboratory Estimates of Trace Gas Emissions Following Surface Application and Injection of Cattle Slurry", *Journal of Environmental Quality*, 29.

Follett, R.F. and J.L. Hatfield (2001), *Nitrogen in the Environment*, Elsevier Press, Amsterdam.

Follett, J.R. et al. (2010), "Environmental and Human Impacts of Reactive Nitrogen", in Delgado J.A. and R.F. Follett (eds.), *Advances in Nitrogen Management for Water Quality*, Soil and Water Conservation Society.

Fox, R.H. et al. (1996), "Estimating Ammonia Volatilization Losses From Urea Fertilizers Using a Simplified Micrometeorological Sampler", *Soil Science Society of America Journal*, 60.

Freney, J.R. et al. (1981), "Ammonia Volatilization", *Ecological Bulletin*, 33.

Galloway, J.N. et al. (2003), "The Nitrogen Cascade", *BioScience*, 53(4).

Graham, R. F et al. (2017), "Comparison of Organic and Integrated Nutrient Management Strategies for Reducing Soil N_2O Emissions", *Sustainability*, 9(4), doi.org/10.3390/su9040510.

Henandez, M. and W. Mitsch (2006), "Influence of Hydrologic Pulses, Fooding Frequency, and Vegetation on Nitrous Oxide Emissions From Created Riparian Marshes", *Wetlands*, 26(3).

Hernandez-Ramirez, G. et al. (2009), "Greenhouse Gas Fluxes in an Eastern Corn Belt Soil: Weather, Nitrogen Source, and Rotation", *Journal of Environmental Quality*, 38.

Hey, D.L. et al. (2005), "Nutrient Farming: The Business of Environmental Management", *Ecological Engineering*, 24.

Hirschi, M. et al. (1997), *60 Ways Farmers Can Protect Surface Water*, North Central Regional Extension Publication 589, Information Technology and Communication Services, College of Agricultural, Consumer and Environmental Sciences, University of Illinois, Urbana-Champaign, IL.

Hutchinson, G.L. et al. (1982), "Ammonia and Amine Emissions from a Large Cattle Feedlot", *Journal of Environmental Quality*, 11.

Iowa Soybean Association (2008), "Understanding Soil Nitrogen Dynamics", *On-Farm Network Update* (October).

Jacobson, L.D. et al. (1999), *Generic Environmental Impact Statement on Animal Agriculture; A Summary of the Literature Related to Air Quality and Odor (H)*, Report prepared for Minnesota Environmental Quality Board, Minnesota Dept. of Agriculture.

Jansson, M. et al. (1994), "Wetlands and Lakes as Nitrogen Traps," *Ambio*, 23(6).

Lam, S. K. et al. (2017), "Using Nitrification Inhibitors to Mitigate Agricultural N$_2$O Emission: a Double-edged Sword?", *Global Change Biology*, 23(2), doi.org/10.1111/gcb.13338.

Legg, J.O. and J.J. Meisinger (1982), "Soil Nitrogen Budgets", in Stevenson, F.J. (ed.), *Nitrogen in Agricultural Soils*, Agronomic Monograph 22, American Society of Agronomy.

Linker, L.C. et al. (2013), "Computing Atmospheric Nutrient Loads to the Chesapeake Bay Watershed and Tidal Waters", *Journal of the American Water Resources Association*, www.chesapeakebay.net/documents/Atmo_Dep__CB_TMDL_10-13.pdf.

MacKenzie, A.F. et al. (1997), "Nitrous Oxide Emission as Affected by Tillage, Corn-soybean-alfalfa Rotations and Nitrogen Fertilization", *Canadian Journal of Soil Science*, 77(2).

Matson, P.A. et al. (1997), "Agricultural Intensification and Ecosystem Properties", *Science*, 277.

Meek, B.D. et al. (1995), « Nitrate Leaching under Furrow Irrigation as Affected by Crop Sequence and Tillage", *J. Soil Sci. Soc. Amer.*, 59.

Meisinger, J.J. and J.A. Delgado (2002), "Principles for Managing Nitrogen Leaching", *Journal of Soil and Water Conservation* 57.

Mitsch, W.J. and J.W. Day (2006), "Restoration of Wetlands in the Mississippi-Ohio-Missouri (MOM) River Basin: Experience and Needed Research", *Ecological Engineering*, 26.

Moore, P.A. et al. (2000), "Reducing Phosphorus Runoff and Inhibiting Ammonia Loss from Manure with Aluminum Sulfate", *Journal of Environmental Quality*, 29(1).

Mosier, A.R. and L. Klemedtsson (1994), "Measuring Denitrification in the Field," in Weaver, R.W. et al. (eds.), *Methods of Soil Analysis: Part 2, Microbiological and Biochemical Properties,* Soil Science Society of America.

National Research Council (2003), *Air Emissions from Animal Feeding Operations: Current Knowledge, Future Needs,* Ad Hoc Committee on Air Emissions from Animal Feeding Operations, Committee on Animal Nutrition, National Research Council, National Academy Press, Washington, D.C.

Oenema, O. et al. (2001), "Gaseous Nitrogen Emissions from Livestock Farming Systems", Chapter 10 in R.F. Follett and J.L. Hatfied (eds.), *Nitrogen in the Environment: Sources, Problems, and Management,* Elsevier Press, Shannon, Ireland.

Peoples, M.B. et al. (1995), "Minimizing Gaseous Losses of Nitrogen", in P.E. Bacon (ed.), *Nitrogen Fertilization in the Environment*, Marcel Dekker, Inc., New York City.

Pellerin, S. et al. (2013), Quelle contribution de l'agriculture française à la réduction des émissions de gaz à effet de serre ? Potentiel d'atténuation et coût de dix actions techniques, Rapport d'étude, Institut national de la recherche agronomique (INRA), France.

Petrolia, D.R. and P.H. Gowda (2006), "Missing the Boat: Midwest Farm Drainage and Gulf of Mexico Hypoxia", *Review of Agricultural Economics*, 28(2).

Rajsic, P. and A. Weersink (2008), "Do Farmers Waste Fertilizer? A Comparison of Ex Post Optimal Nitrogen Rates and Ex Ante Recommendations by Model, Site and Year", *Agricultural Systems*, 97(1-2).

Randall, G.W. et al. (2008), "Nitrogen Management To Protect Water Resources", in Schepers, J.S. et al. (eds.), *Nitrogen in Agricultural Systems*, Agronomy Monograph 49, Soil Science Society of America.

Randall, G.W. et al. (2010), "Nitrogen and Drainage Management To Reduce Nitrate Losses to Subsurface Drainage", in Delgado J.A. and R.F. Follett (eds.), *Advances in Nitrogen Management for Water Quality*, Soil and Water Conservation Society.

Ribaudo, M. et al. (2003), *Manure Management for Water Quality: Costs to Animal Feeding Operations of Applying Manure Nutrients to Land,* Agricultural Economic Report N° 824, Economic Research Service, U.S. Department of Agriculture, www.ers.usda.gov/publications/aer824.

Ribaudo, M. et al. (2011), *Nitrogen in Agricultural Systems: Implications for Conservation Policy,* Economic Research Report, ERR-127, Economic Research Service, U.S. Department of Agriculture, www.ers.usda.gov/webdocs/publications/44918/6767_err127.pdf?v=41056.

Rochette, P. et al. (2004), "Carbon Dioxide and Nitrous Oxide Emissions Following Fall and Spring Applications of Pig Slurry to an Agricultural Soil", *Soil Science Society of America Journal*, 68.

Schepers, J.S. et al. (1986), "Effect of Yield Goal and Residual Soil Nitrogen Considerations on Nitrogen Fertilizer Recommendations for Irrigated Maize in Nebraska", *Journal of Fertilizer Issues*, 3.

Schmaltz, R. (2017), "What is Precision Agriculture?", AgFunderNews, 24 April 2017, agfundernews.com/what-is-precision-agriculture.html.

Schimmelpfennig, D. (2016), *Farm Profits and Adoption of Precision Agriculture*, ERS, October 2016 Economic Research Report, ERR-217, Economic Research Service, U.S. Department of Agriculture, www.ers.usda.gov/webdocs/publications/80326/err-217.pdf?v=42661.

Shaffer, M.J. et al. (2010), "Simulation Processes for the Nitrogen Loss and Environmental Assessment Package (NLEAP)", in J.A. Delgado and R.F. Follett (eds.), *Advances in Nitrogen Management for Water Quality*, Soil and Water Conservation Society.

Sharpe, R.R. and L.A. Harper (1995), "Soil, Plant, and Atmospheric Conditions as They Relate to Ammonia Volatilization", *Fertilizer Research*, 42.

Sheriff, G. (2005), "Efficient Waste? Why Farmers Over-Apply Nutrients and the Implications for Policy Design," *Review of Agricultural Economics* 27(4):542-557.

Shipitalo, M.J. et al. (2000), "Conservation Tillage and Macropore Factors that Affect Water Movement and the Fate of Chemicals", *Soil and Tillage Research*, 53(3-4).

Smith, R.A. et al. (1997), "SPARROW Surface Water-Quality Modeling Nutrients in Watersheds of the Conterminous United States: Model Predictions for Total Nitrogen (TN) and Total Phosphorus (TP)", water.usgs.gov/nawqa/sparrow/wrr97/results.html.

Sweeten, J. et al. (2000), *Air Quality Research and Technology Transfer White Paper and Recommendations for Concentrated Animal Feeding Operations,* Report for US Department of Agriculture, Agricultural Air Quality Task Force, Washington D.C.

Turner, R.E. and N.N. Rabalais (2003), "Linking Landscape and Water Quality in the Mississippi River Basin for 200 Years", *BioScience*, 53(6).

USEPA (2003), *National Management Measures to Control Nonpoint Source Pollution from Agriculture*, U.S. Environmental Protection Agency, Office of Water, EPA 841-B-03-004.

USEPA (2010a), *Inventory of U.S. Greenhouse Gas Emissions and Sinks: 1990-2008,* U.S. Environmental Protection Agency, EPA 430-R-10-006.

USEPA (2010b), *Methane and Nitrous Oxide Emissions from Natural Sources*, U.S. Environmental Protection Agency, Office of Atmospheric Programs, EPA 430-R-10-001.

USEPA-SAB (2011), *Reactive Nitrogen in the United States: An Analysis of Inputs, Flows, Consequences and Management Options*, U.S. Environmental Protection Agency's Science Advisory Board, EPA-SAB-11-013.

USDA-NRCS (2006), "Natural Resources Conservation Service Conservation Practice Standard: Nutrient Management", *National Handbook of Conservation Practice*, U.S. Department of Agriculture, Natural Resource Conservation Service.

USDA-NRCS (2011a), *Assessment of the Effects of Conservation Practices on Cultivated Cropland in the Ohio-Tennessee River Basin*, Conservation Effects Assessment Project, U.S. Department of Agriculture, Natural Resource Conservation Service, www.nrcs.usda.gov/Internet/FSE_DOCUMENTS/stelprdb1046342.pdf.

USDA-NRCS (2011b), *Assessment of the Effects of Conservation Practices on Cultivated Cropland in the Chesapeake Bay Region*, Conservation Effects Assessment Project, U.S. Department of Agriculture, Natural Resource Conservation Service, www.nrcs.usda.gov/Internet/FSE_DOCUMENTS/stelprdb1042076.pdf.

USDA-NRCS (2012a), *Assessment of the Effects of Conservation Practices on Cultivated Cropland in the Great Lakes Region*, Conservation Effects Assessment Project, U.S. Department of Agriculture, Natural Resource Conservation Service, www.nrcs.usda.gov/wps/portal/nrcs/detail/national/technical/nra/ceap/?cid=stelprdb1045403.

USDA-NRCS (2012b), *Assessment of the Effects of Conservation Practices on Cultivated Cropland in the Missouri River Basin*, Conservation Effects Assessment Project, U.S. Department of Agriculture, Natural Resource Conservation Service, www.nrcs.usda.gov/Internet/FSE_DOCUMENTS/stelprdb1048710.pdf.

USDA-NRCS (2012c), *Assessment of the Effects of Conservation Practices on Cultivated Cropland in the Upper Mississippi River Basin*, Conservation Effects Assessment Project, U.S. Department of Agriculture, Natural Resource Conservation Service, www.nrcs.usda.gov/Internet/FSE_DOCUMENTS/stelprdb1042093.pdf.

USDA-NRCS (2013a), *Assessment of the Effects of Conservation Practices on Cultivated Cropland in the Arkansas-White-Red River Basin*, Conservation Effects Assessment Project, U.S. Department of Agriculture, Natural Resource Conservation Service, www.nrcs.usda.gov/Internet/FSE_DOCUMENTS/stelprdb1088485.pdf.

USDA-NRCS (2013b), *Assessment of the Effects of Conservation Practices on Cultivated Cropland in the Lower Mississippi River Basin*, Conservation Effects Assessment Project, U.S. Department of Agriculture, Natural Resource Conservation Service, www.nrcs.usda.gov/Internet/FSE_DOCUMENTS/stelprdb1176978.pdf.

USDA-NRCS (2013c), *Impacts of Conservation Adoption on Cultivated Acres of Cropland in the Chesapeake Bay Region, 2003-06 to 2011*, Conservation Effects Assessment Project, U.S. Department of Agriculture, Natural Resource Conservation Service, www.nrcs.usda.gov/wps/portal/nrcs/detail/national/technical/nra/ceap/na/?cid=stelprdb1240074.

USDA-NRCS (2014a), *Assessment of the Effects of Conservation Practices on Cultivated Cropland in the Souris-Red-Rainy Basin*, Conservation Effects Assessment Project, U.S. Department of Agriculture, Natural Resource Conservation Service, www.nrcs.usda.gov/Internet/FSE_DOCUMENTS/stelprdb1260130.pdf .

USDA-NRCS (2014b), *Assessment of the Effects of Conservation Practices on Cultivated Cropland in the South Atlantic Gulf Basin*, Conservation Effects Assessment Project, U.S. Department of Agriculture, Natural Resource Conservation Service, www.nrcs.usda.gov/Internet/FSE_DOCUMENTS/stelprdb1256674.pdf .

USDA-NRCS (2014c), *Assessment of the Effects of Conservation Practices on Cultivated Cropland in the Delaware River Basin*, Conservation Effects Assessment Project, U.S. Department of Agriculture, Natural Resource Conservation Service, www.nrcs.usda.gov/Internet/FSE_DOCUMENTS/stelprdb1263627.pdf.

USDA-NRCS (2014d), *Assessment of the Effects of Conservation Practices on Cultivated Cropland in the Pacific Northwest Basin*, Conservation Effects Assessment Project, U.S. Department of Agriculture, Natural Resource Conservation Service, www.nrcs.usda.gov/Internet/FSE_DOCUMENTS/stelprdb1256682.pdf.

USDA-NRCS (2015), *Assessment of the Effects of Conservation Practices on Cultivated Cropland in the Texas Gulf Basin*, Conservation Effects Assessment Project, U.S. Department of Agriculture, Natural Resource Conservation Service, www.nrcs.usda.gov/Internet/FSE_DOCUMENTS/nrcseprd374812.pdf.

Vitousek, P.M., et al. (1997), "Human Alteration of the Global Nitrogen Cycle: Sources and Consequences," *Ecological Applications,* 73(3).

Wolfert, S. et al. (2017), "Big Data in Smart Farming – A Review", *Agricultural Systems*, 153, doi.org/10.1016/j.agsy.2017.01.023.

Wulf, S. et al. (2002), "Application Technique and Slurry Co-Fermentation Effects on Ammonia, Nitrous Oxide, and Methane Emissions After Spreading: II Greenhouse Gas Emissions", *Journal of Environmental Quality*, 31.

Zilberman, D. et al. (2001), "Innovative Policies for Addressing Livestock Waste Problems", White Paper Summaries, National Centre for Manure and Animal Waste Management.

Chapter 5. Criteria to guide nitrogen policy making

This chapter provides a framework for analysing the merits of nitrogen management policy instruments. It establishes a typology of the different types of instruments available to decision makers and proposes three criteria to evaluate them (effectiveness, cost-efficiency and feasibility). It stresses the importance of strengthening the coherence between nitrogen pollution management policy and other policies, both environmental and sectoral.

Whatever the policy approach is (risk or precautionary), evaluation criteria are needed to select the right risk management or uncertainty management tools (Figure 5.1). Firstly, it is necessary to evaluate and address the unintended effects on nitrogen of sectoral (e.g. agricultural policy, energy policy) and environmental policies (climate policy and others). Nitrogen policy instruments that are cost-effective and whose "feasibility" of implementation is not problematic may then be selected. Finally, due to the nitrogen cascade, the unintended effects of an instrument targeting one form of nitrogen on other forms of nitrogen should be estimated, in order to promote synergies and avoid "pollution swapping".

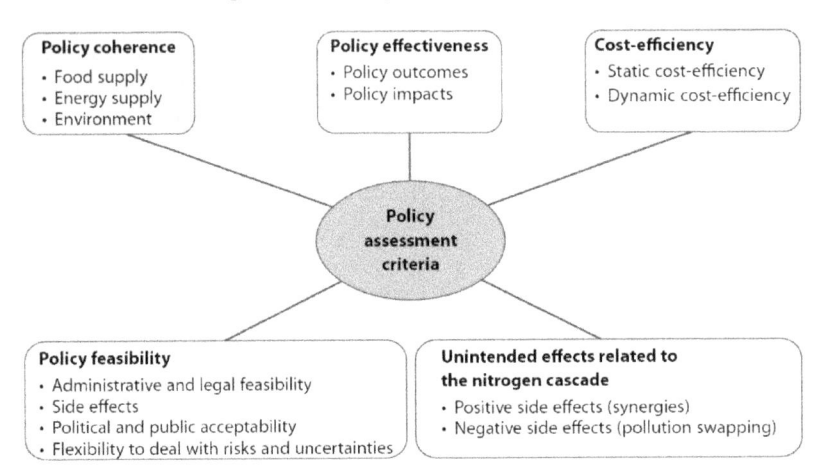

Figure 5.1. Policy assessment criteria

Policy coherence
- Food supply
- Energy supply
- Environment

Policy effectiveness
- Policy outcomes
- Policy impacts

Cost-efficiency
- Static cost-efficiency
- Dynamic cost-efficiency

Policy feasibility
- Administrative and legal feasibility
- Side effects
- Political and public acceptability
- Flexibility to deal with risks and uncertainties

Unintended effects related to the nitrogen cascade
- Positive side effects (synergies)
- Negative side effects (pollution swapping)

Note: The policy coherence criteria applies to policies not primarily focused on the management of nitrogen pollution.
Source: Adapted from Drummond et al. (2015).

An example of what the search for consistency with agricultural policy may involve is given in Section 5.1 below.[1] Section 5.2 introduces the criteria of effectiveness, cost-efficiency and feasibility for designing nitrogen policy instruments. Applying these criteria to different types of policy instruments, Chapter 6 presents their advantages and disadvantages, in a generic sense, as well as case studies on nitrogen policy instruments. The assessment of the unintended effects of the nitrogen cascade is just beginning (see Section 5.3); Chapter 4. presented a case study on agricultural practices.

5.1 Policy coherence

First and foremost, it must be ensured that sectoral policies do not encourage excessive nitrogen production. This may be the case with some policies aimed at stimulating agricultural production or the security of energy supply.

For example, Bartelings et al., 2016 estimated the impact of subsidies on fertiliser production and use on greenhouse gas (GHG) emissions using a computable general equilibrium model.[2] First, ad-valorem subsidy rates were calculated from information on fertiliser costs. For example, Indonesia and India have a policy of reducing energy costs in fertiliser production in order to provide fertiliser to national farmers at a lower cost.[3] It is estimated that the resulting public financial support reduces the cost of nitrogen fertiliser for Indonesian farmers by 68%

below production costs and by 56% for Indian farmers (Table 5.1). Russia and China support fertiliser use through direct subsidies to farmers. It is estimated that the input subsidy (implemented as an area payment in China) is equivalent to an ad valorem subsidy of about 28% of the value of fertiliser cost in Russia, and 12.5% in China (Table 5.1).

Table 5.1. Fertiliser subsidies in selected BRIICS countries

Ad valorem (% of fertiliser cost)	Input subsidy to farmers	Support to fertiliser producers
Indonesia		68
India		56
Russia	28	
China	12.5	

Source: Von Lampe et al. (2014).

Second, GHG emissions in relation to fertiliser production and use were estimated (Table 5.2). This is obviously an approximation since nitrous oxide (N_2O) emissions from crops depend on farming practices and not just on fertilisation rates. In particular, the application of the 4R concept (the right type of fertiliser at the right rate applied at the right time in the right place)[4] can both increase the efficiency of nitrogen uptake by plants and reduce excess nitrogen in the field, thus reducing N_2O emissions (Omonode et al., 2017).

Table 5.2. Greenhouse gas (GHG) emissions related to use and production of nitrogen fertilisers

GHG emission (g/kg)	Carbon dioxide (CO_2)	Nitrous oxide (N_2O)	Methane (CH_4)	CO_2eq[1]
Fertiliser production	2 827	10	9	5 729
Fertiliser use	0	21	0	5 565

Note:
1. The CO_2 equivalent, abbreviated CO_2-eq, is used to compare the emissions of various GHGs according to their global warming potential in relation to CO_2 over the next hundred years (GWP_{100}). CO_2-eq for a GHG is obtained by multiplying the quantity emitted by the associated GWP_{100}. For the purpose of this report, GWP_{100} of 265 for N_2O and 28 for CH_4 were used based on the Fifth Assessment Report (AR5) of the Intergovernmental Panel on Climate Change (www.ipcc.ch/pdf/assessmentreport/ar5/wg1/WG1AR5_Chapter08_FINAL.pdf). GWP values for non-CO_2 gases do not include climate-carbon feedbacks.
Source: Adapted from Bartelings et al. (2016).

GHG emissions of crops also depend on the type of fertiliser (with N_2O emissions typically being higher for urea than for ammonium nitrate) and soil type, with N_2O emissions generally high for clay soils with poor drainage (Figure 5.2). Previous internationally accepted estimates were that for every kg of nitrogen fertiliser applied in grain crop production, there is a loss of 1% as N_2O to the atmosphere.[5]

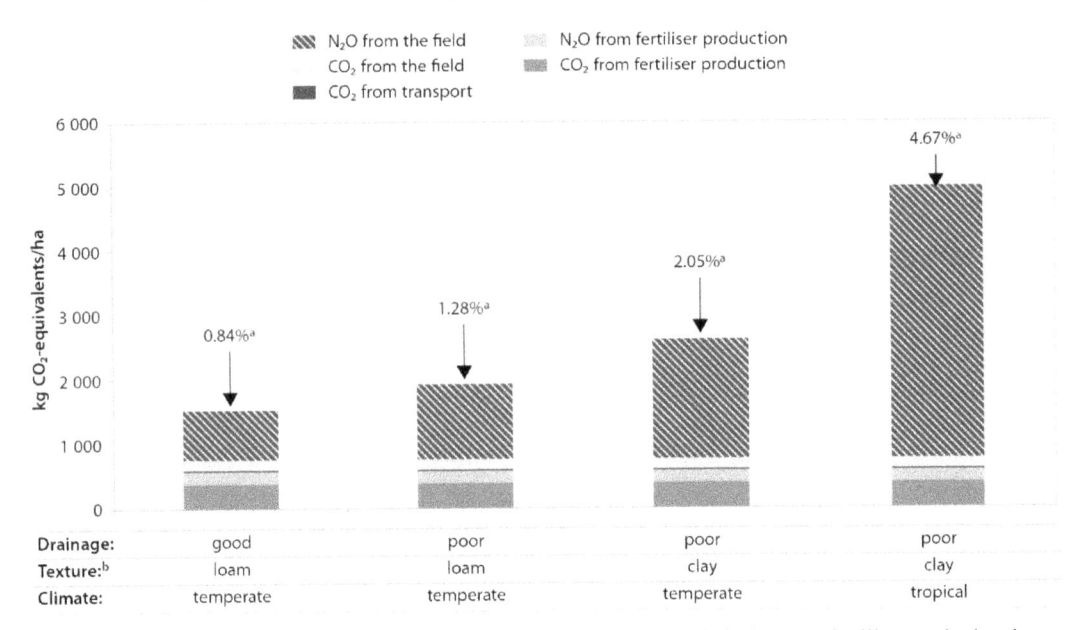

Figure 5.2. Greenhouse gas (GHG) emissions of crops by soil type

a) Nitrous oxide (N₂O) emissions from the field as a % of applied nitrogen fertiliser; calculated according to the Bouwman model (Bouwman et al., 2002) with an uncertainty range of -40% to + 70%.
b) Other important soil factors are soil organic carbon and pH.
Source: Brentrup and Pallière (2009).

Finally, the GHG emission reduction impact of phasing out fertiliser subsidies was estimated. The impact differs by country. It is particularly pronounced in India and Indonesia where, with the elimination of subsidies paid directly to the fertiliser industry, locally produced nitrogen would become less attractive than imported phosphorus and potassium (substitution effect between the three nutrients). The model shows that GHG emissions would also decrease in China, where the removal of input-related area payments (linked to both land and fertiliser use) would lead to a reduction in agricultural production. While GHG emissions from nitrogen use would fall in Russia, emissions from fertiliser production would rise as Russia would begin to export nitrogen to India and Indonesia.

According to the study, the abolition of fertiliser subsidies would only have a modest effect on agricultural land use in all four countries, except for China. In China, the phasing out of fertiliser subsidies could lead to an increase in carbon sinks, as forests could develop on land that is no longer used for agricultural purposes, adding to the net reduction of GHG emissions.

In fact, China has taken steps to phase out fertiliser subsidies and aims to cap fertiliser use by 2020. The 2020 Zero-Growth Action Plan for Chemical Fertilisers and Pesticides aims to reduce the annual growth of chemical fertiliser use to below 1% for the 2015-19 period and achieve zero-growth by 2020 for major agricultural crops (OECD, 2016b). However, these policy developments need to be assessed in a context of increasing policy support for Chinese farmers, largely through forms of support that distort agricultural production (Figure 5.3).

Figure 5.3. China and OECD: trends in level and structure of agricultural support

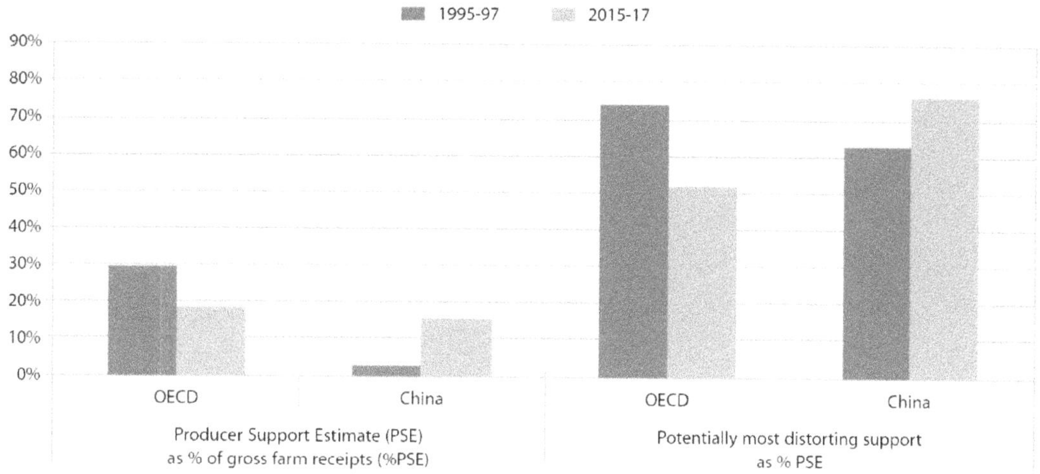

Source: OECD (2018a).

To sum up, some forms of agricultural support can distort input use and agricultural production and thus have negative environmental impacts (such as increased emissions of N_2O). Before designing targeted policies on nitrogen pollution, it is therefore essential to monitor and evaluate sectoral policies and their unwanted effects on nitrogen emissions. The OECD has established a typology of support measures for agricultural producers and is working on an assessment of their environmental impact (see for example OECD, 2018b). The OECD has also developed expertise in the monitoring and evaluation of policies that directly support the production or consumption of fossil fuels. This OECD work could provide a basis for assessing the adverse effects of sectoral policies on nitrogen pollutant emissions.

Policy coherence must also be sought with environmental policies that are not primarily aimed at reducing nitrogen pollution. For example, in New Zealand, a policy to reduce carbon dioxide (CO_2) emissions has helped reduce the leaching of nitrate (NO_3^-) into water (in part, following OECD, 2013). A GHG Emissions Trading Scheme (ETS) allows GHG emitters who do not want to reduce their emissions to enter into agreements (through a Trust) with farmers who agree to sequester carbon. With financial compensation, farmers undertake to convert pastoral lands into forests, thereby helping to reduce the leaching of NO_3^- into water. GHG emitters receive ETS credits in exchange for pastoral land converted to forestry (Figure 5.4).

Figure 5.4. New Zealand's greenhouse gas (GHG) Emissions Trading Scheme

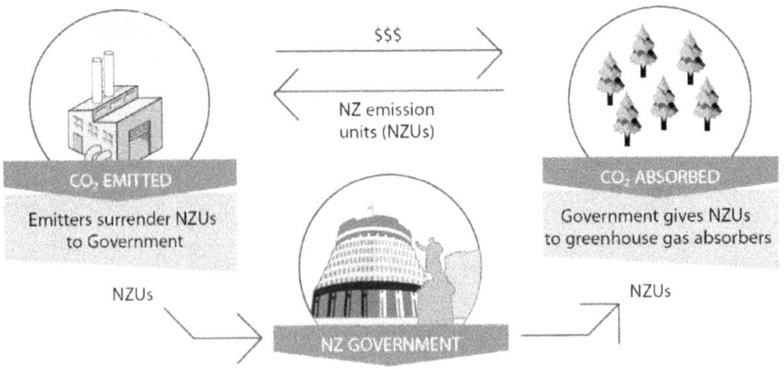

Source: Ministry for the Environment of New Zealand, www.mfe.govt.nz/climate-change/reducing-greenhouse-gas-emissions/about-nz-emissions-trading-scheme, accessed 29 March 2018.

This is occurring in Lake Taupo, the largest lake in the country, which is threatened by nitrogen pollution. To reduce nitrogen pollution in the lake, a cap-and-trade system for nitrogen was put in place for farmers around Lake Taupo. Instead of selling their nitrogen pollution rights, farmers can opt for permanent reductions in nitrogen, for which they are compensated financially through the Trust. In turn, the Trust is funded by GHG emitters through the purchase of ETS credits. Farmers are paid for the reduction of nitrogen emissions, at the same time they receive income from forest credits.

This example of environmental measures not focusing on nitrogen pollution but helping to manage it explains why coherence must also be sought in environmental policies. As a counterexample, assessing trade-offs between climate change mitigation goals and nitrogen pollution management may be more challenging when it comes to forest fertilisation,[6] which increases both the sequestration of carbon and the risks associated with nitrogen. According to de Vries, 2017 nitrogen deposition has contributed to a significant increase in carbon sequestration in forests and forest soils in Europe since 1950. However, nitrogen aerosol deposition may also contribute to decreasing the drought tolerance of forest trees by decreasing the opening of stomata[7] in their leaves (Burkhardt and S.Pariyar, 2014; Grantz et al., 2018).

5.2 The effectiveness, efficiency and feasibility of policy instruments[8]

5.2.1 A typology of policy instruments

Policy instruments of extremely varied design and practical application may be employed for the control of nitrogen pollution. Based on the classifications provided by OECD (2008), such instruments may be placed into the following categories:

- Pricing instruments: environmentally related taxes and charges and tradable permit systems (TPS)
- Direct environmental regulation (DER)

- Financial support instruments: public financial support (PFS) and payments for ecosystem services (PES)
- Information measures
- Voluntary schemes

Environmentally related taxes are compulsory duties levied by a competent authority (either a national, regional or local government) on legal entities concerning certain products, activities or transactions with the aim of discouraging environmentally harmful behaviour through price disincentives. Environmental charges are payments made by consumers to providers of environmental management services, such as wastewater treatment.

The TPS are quantity-based instruments that assign property rights to units of positive or negative (actual or potential) externalities (such as pollutants or ecosystem services), which after allocation may be traded as a commodity between obligated entities on newly-created markets. The most common design is a 'cap-and-trade' approach, in which an aggregate cap on the volume of a pollutant is set (below that which would be produced in absence of a cap) and permits assigned for each unit of pollutant under the cap, affording the bearer a 'right' to emit or discharge a unit of pollutant, creating scarcity and a market value. 'Baseline and credit' systems, rather than setting a cap with a finite number of tradable permits, create and award credits (permits) to entities that reduce pollution levels below a projected baseline, which may then be traded as above (with an aggregate cap retained and set by the projected baseline). A similar approach may be taken concerning ecosystem services i.e. credits issued for protecting or restoring ecosystem services, which may be traded to other entities to offset an increase in the need for such services – for example, reforestation to increase CO_2 sequestration to offset increasing CO_2 emissions; or creating wetlands to maximize denitrification in NO_3^- enriched watersheds (Hansen et al., 2016).

DER refer to the imposition of minimum or maximum standards, limits or bans on, or requirements for, certain products, practices or performance levels (such as 'Best Available Technologies' (BAT)), and other similar obligations, such as activity permits.

PFS refers to the use of the public budget to encourage practices with reduced environmental impact (or discourage those with high impact), to foster organisational or technological innovation, or to finance infrastructure. Such support may be given via numerous channels, including direct budget allocations or grants, low or zero interest loans, loan guarantees, or preferential tax treatment (OECD, 2008).

Various definitions of the PES concept exist, with no agreed single definition (Schomers and Matzdorf, 2013). However, PES may be broadly defined as a transaction where ecosystem managers (e.g. land owners) are compensated through conditional payments by the beneficiaries of ecosystem services (including governments, where the public is the general beneficiary) for the additional cost of maintaining ecosystem services above legally required levels.

Information measures attempt to overcome information failures through awareness campaigns, labelling and certification schemes, training and education programmes, the provision of 'best practice' guidelines, and disclosure and reporting mechanisms.

Voluntary schemes include environmental management systems, voluntary agreements negotiated between government and certain economic sectors, unilateral sectoral or cross-sectoral agreements to achieve a given environmental objective.

No single instrument or type of instrument is appropriate for tackling all sources of pollution, from all sectors. Whilst this section introduces types of policy instruments individually, it will probably be desirable to introduce instruments in combination, as a policy 'mix'. Both individual instruments and policy mixes experience trade-offs between the three assessment criteria discussed below. Such trade-offs are unavoidable in a 'second-best' world, characterised by multiple market failures and uncertainties.

5.2.2 Effectiveness, efficiency and feasibility criteria

Existing policy analysis methodologies tend to assess the performance of a policy instrument against the criteria of economic efficiency – particularly static efficiency (i.e. the most productive balance between resource allocation and outcome, or when a desired outcome is given, achievement of this outcome at the lowest cost at a given point in time (static cost-efficiency)). Such analyses often recommend the use of pricing tools as cost-minimising solutions to reduce the emission of pollutants, particularly those encompassing all (anthropogenic) sources of the pollutant, over the full geographic scope in question. However, such a view neglects 'real world' complications that often act to limit policy effectiveness and the ability to introduce instruments in the more theoretically desirable manner. Such factors include (Görlach, 2013).

- Market failures aside from the externality of the pollutant in question. Examples are information failures and principal-agent problems (split incentives), along with 'satisficing' behaviour,[9] which can act to prevent the market from producing an 'optimal' outcome.

- The development of abatement options. This may be stimulated to some extent by pricing policies, however such development is uncertain, and a high price may be required to stimulate private investment in the development of such technologies and practices.

- Administrative feasibility, including pollutant and compliance monitoring, and enforcement. For some sources of pollution it may not be possible, or prohibitively expensive, to administer certain instruments. Any efficiency gain associated with a pricing instrument may be more than offset by the transaction costs of administration, monitoring and ensuring compliance for such sources.

- Lack of public and political acceptability, which, along with legal compatibility, may prevent an otherwise suitable instrument from being introduced.

- Distributional impacts and equity issues, which may also prevent an instrument's introduction if they cannot be adequately addressed.

- The global nature of some pollutants, such as GHGs. Policy instruments that simply shift such pollution to another jurisdiction rather than producing global abatement cannot be considered truly effective.

Such conditions suggest that broader policy assessment criteria are required to capture these nuances, and allow for a more rounded evaluation of the relative desirability of policy instruments for different applications.

'Effectiveness' concerns the results of a policy intervention, which may be defined in terms of policy outcomes (e.g. the use of catalytic converters in vehicles) or policy impacts (e.g. reduced disturbance of the nitrogen cycle). Which of these is applicable depends on the specific objectives of a given policy instrument, and of the assessment being carried out. For example, whilst the ultimate objective for instruments aiming to control nitrogen pollution is reduced disturbance of the natural nitrogen cycle, it may be impractical to ascribe such an impact to a single policy intervention. In such cases, policy objectives (if made explicit) are often defined in terms of a policy 'outcome' (Görlach, 2013).

Broadly, the 'cost-efficiency' criterion relates the inputs (costs) to the results (outcome/impact) of a policy intervention, and asks whether the intervention's objective is being achieved at the least cost to society. This is composed of two sub-criteria. The first is static cost-efficiency, which requires all targeted sources of a given pollutant to face the same marginal abatement cost, and therefore equal incentive for abatement, so that any level of abatement from these sources is reached at the lowest cost for society as a whole, using the abatement options available at a given point in time. This is sometimes referred to as the equi-marginal principle. The second sub-criterion is dynamic cost-efficiency, which refers to achieving pollution abatement at the lowest possible societal cost over a given time period, by providing a continuous incentive to innovate and implement ever-cheaper abatement options (Duval, 2008). As such, dynamic cost-efficiency assumes temporally flexible rather than fixed abatement options and costs. From this perspective, encouraging the deployment of a currently expensive abatement technology to enable innovation and produce low-cost abatement for the future, in contrast to a static perspective, may be preferable to a lower cost-alternative with less potential for long-term cost-reduction (Görlach, 2013).

Pricing instruments are statically cost-efficient if the explicit or implicit price is applicable to and equalised across all sources of pollution within the scope of the instrument. A tax or TPS may have a broad geographical and sectoral scope, and may cover multiple fungible 'commodities' (pollutants). For example, pollutants that have the same (or comparable) impact, such as the production of eutrophication of water bodies by nitrogen and phosphorus. Pollutants that have different but stable levels of impact in relation to each other, such as GHGs with different global warming potentials (GWP), may also be considered in a single tradable permit system. For example, the number of permits required for the emission of a unit of a GHG with a high GWP may increase in proportion to the number of permits required for the emissions of a unit of a GHG with the lowest GWP within the scope of the instrument.

The 'feasibility' of instrument implementation is obviously essential if an instrument is actually to be introduced and function effectively. The concept may be broken down into five general aspects. The first is *administrative feasibility*, including, for example, the ease of administration, transaction costs and stringency of the compliance regime. The second concerns the ability to address *side effects*, intended and unintended, positive and negative. This may include negative distributional/equity impacts and impacts on industrial competitiveness, but also job creation, resource security and health and environmental co-benefits (Box 5.1). Linked to these two aspects are *political and public acceptability*, which are themselves clearly linked. The fourth aspect is *legal and institutional feasibility*, which considers the compatibility of a policy instrument with existing legal frameworks and constitutional doctrines. The final aspect concerns the *flexibility* of an instrument, and its ability to respond to new information and deal with risks and uncertainties. This also reflects the ability of an instrument to deal with strategic behaviour (either induced by asymmetric information, rent-seeking, regulatory capture, or even fraud) (Görlach, 2013).

Box 5.1. Example of positive side effect of environmental policies primarily aimed at reducing nitrogen pollution

The activities causing agricultural emissions of nitrogen and GHGs to a large extent overlap. When the agricultural emission of nitrogen is reduced, it will therefore also lead to a reduction of the emissions of GHGs from agriculture. The EU Water Framework Directive (2000/60/EC) sets a target of good ecological status for all surface waters and good status for groundwater which should be achieved no later than 2027. In Denmark this was translated into a set of geographically differentiated reduction targets for nitrogen loads to coastal waters and groundwater. The fulfilment of these nitrogen targets is projected to reduce the agricultural emission of GHGs by up to two million tonnes by 2027.

Under current commitments, Denmark must reduce emissions of GHGs in sectors of the economy that fall outside the scope of the EU Emissions Trading System (ETS) by 39% in 2030, compared to 2005. Based on a projection of emissions by the Danish Energy Agency from 2017, it is estimated that emissions in the year 2030 must be reduced by 2.5 million tonnes in order to reach this target. Fulfilment of the EU Water Framework Directive can therefore imply that there is only a modest need of further reductions in the Danish non-ETS sector in order to reach the target. Even though the estimate of the effect of fulfilling the Water Framework Directive is quite uncertain the results illustrate that the level of nitrogen regulation can have a high impact on the total emissions of GHGs.

Source: Danish Economic Councils (2018).

For example, Jacobsen et al., 2017 argue that the legal and regulatory complexity of adopting mandatory agricultural (land use related) measures at the national level to achieve site-specific environmental objectives was underestimated in a top-down political process in the context of implementing the EU Water Framework Directive in Denmark. The ambitious mandatory policy measures, which added to existing high regulation pressure, led to regulatory challenges, such as possible violation of private property rights. Consequently, the political acceptability and legitimacy of the measures were undermined, resulting in their gradual withdrawal. It is argued that the adoption of more flexible measures to be implemented at the local level could have resulted in fewer difficulties from an economic and legal point of view as measures could have been applied where there was a clear environmental benefit, and possibly also at a lower cost.

Generally, environmentally related taxation and charging instruments are highly administratively feasible; most countries have the required institutions and administrative systems already in place. Even so, there are various exceptions and nuances. A tax, charge or tradable permit system is likely to be more administratively feasible than downstream application, as the former targets far fewer actors than the latter, reducing transaction costs and increasing the potential for effective monitoring and enforcement. Furthermore, by its nature, the potential for avoidance and evasion is reduced (Matthews, 2010). However, technical challenges are present for both upstream and downstream focused instruments. Whilst pollutants such as CO_2 and sulphur dioxide (SO_2) emissions depend largely on the carbon or sulphur content of the fuel, nitrogen oxides (NO_x) and N_2O emissions from the combustion process depends largely on the combustion technology employed, rendering an efficient upstream pricing instrument infeasible. Conversely, monitoring nitrogen pollution at the point of emission from small stationary or mobile combustion sources (e.g. households and transport), or from the multiple and diffuse pathways of non-point source pollution in the agriculture sector, may be both technically and administratively infeasible. For such sources, a pricing instrument applied to ambient conditions (e.g. pollutant concentration in a water body), may be a practical solution. However, such an approach may only be effective if individual actors believe that their emissions substantially impact the aggregate, producing an incentive to reduce emissions. This may be the case concerning water pollution in a small watershed with few agricultural producers, but not, for example, in an urban area with large numbers of vehicles entering, leaving and individually contributing very marginally to aggregate pollution levels (Karp, 2005).

5.3 Unintended effects related to the nitrogen cascade

Beyond "generic" criteria such as policy coherence, on the one hand, and the effectiveness, efficiency and feasibility of nitrogen policy instruments, on the other hand, it is crucial to consider an additional criterion to guide nitrogen policy making. This last criterion takes into account the reality of the nitrogen cascade, that is to say that once fixed and due to its labile nature, nitrogen tends to change form until it eventually reverts to dinitrogen (see Chapter 1. for the description of the cascade). The objective is to refine the design of nitrogen policy instruments (compared to a design based on the first two criteria alone) by evaluating their unintended effects on other forms of nitrogen due to the nitrogen cascade.

Whatever the policy approach is (risk, precautionary), assessing nitrogen policy instruments should consider the ancillary effects on other nitrogen forms, both positive and negative. In particular, efforts to lessen the impacts caused by nitrogen in one area of the environment should (i) not result in unintended nitrogen impacts in other areas ("pollution swapping" effects)[10], and (ii) seize opportunities to reduce other nitrogen impacts ("synergy" effects). It is therefore necessary to assess risk-risk trade-offs[11] of various policies or best management practices,[12] whether in agriculture, fossil fuel combustion, industrial processes or treatment of wastewater.

For example, the use of selective catalytic reduction (SCR) systems to abate vehicular NO_x emissions brings new concerns on the emissions of the byproducts ammonia (NH_3) and N_2O (even though overall nitrogen emissions are greatly reduced). Since the SCR system (which is a urea-based $deNO_x$ system)[13] is used to reduce NO_x emissions, more urea solution would need to be injected in the SCR to react and reduce NO_x emissions. As a consequence, higher emissions of N_2O and NH_3 could be expected (Suarez-Bertoa et al., 2016).

By contrast, tertiary treatment of sewage to remove NO_3^- allows drastically reducing – by more than 90% – N_2O emissions from sewage relative to the absence of such treatment (Box 5.2). However, such tertiary treatment leads to the production of sludge, the disposal of which by incineration (as is the norm in Switzerland, for example) releases NO_x.

Box 5.2. The synergy effect on nitrous oxide (N_2O) emissions of practices to remove nitrate (NO_3^-) from sewage

Tertiary treatment to remove nitrogen from sewage involves a two-step process of nitrification and denitrification (often carried out in separate areas in the wastewater treatment plant (WWTP) as the first requires aerobic conditions and the second anaerobic conditions). N_2O can by formed during both the nitrification and denitrification stages.[14] A survey in 2011 revealed a large variation in N_2O emission among the WWTPs surveyed (STOWA GWRC, 2011). This is because N_2O emissions depend on the type and design of the WWTP (e.g. aeration mode) and the frequency of changes in process conditions[15] (Kampschreuer et al., 2009).

The Intergovernmental Panel on Climate Change published guidelines to estimate N_2O emissions from WWTPs, as part of national GHG inventories. The guidelines distinguish "minimal nitrogen removal during treatment" and "controlled nitrification and denitrification steps".

The first case assumes that all nitrogen entering the WWTP is discharged into recipient water bodies, where it is mineralised, nitrified and denitrified by natural processes. During these processes some of the discharged nitrogen will be emitted as N_2O, at a default factor of 0.005 kg $N-N_2O$/kg N discharged, with an uncertainty range of 0.0005 to 0.25 kg $N-N_2O$/kg N discharged (IPCC, 2006).

The proposed default emission factor with "controlled nitrification and denitrification steps" is 0.0032 kg N_2O/person/year (0.0020 kg $N-N_2O$/person/year), with a range of uncertainty from 0.002 to 0.08 kg $N-N_2O$/person/year. Assuming sewage nitrogen loading of 16 g N/person/day for developed countries, this equates to approximatively 0.00035 kg $N-N_2O$/kg N discharged (Foley and Lant, 2009).

Consequently, tertiary treatment in WWTPs reduces the amount of N_2O released from an emission factor of 0.005 kg $N-N_2O$/kg N discharged to an emission factor of 0.00035 kg $N-N_2O$/kg N treated (i.e. 93% of N_2O emissions avoided).

Bacterial activity can open up new synergy possibilities for removing N_2O and other nitrogen forms (i.e. to convert nitrogen to dinitrogen) through denitrification. For example, the use of anammox bacteria for wastewater treatment would prevent N_2O emissions. Anammox (an abbreviation for ANaerobic AMMonium OXidation) is an alternative denitrifying pathway which occurs in waters (and sediments) that are naturally low in oxygen (anoxic areas).[16] With the discovery of anammox (1999), scientists showed that some bacteria can draw their energy from nitrite (NO_2^-) and ammonium (NH_4^+) – instead of NO_3^-, returning them back directly into dinitrogen (i.e. completely skipping N_2O). The use of anammox bacteria for wastewater treatment could also be cost-efficient since pumping oxygen into water represents half of the operating costs of a wastewater treatment plant (Lawson et al., 2017). An additional advantage of anammox bacteria, compared to conventional wastewater treatment, is that they convert a larger amount of NH_4^+ to dinitrogen. A main implementation challenge, though, is that anammox bacteria grow very slowly.

In addition to benefits on biodiversity and carbon storage, wetland restoration may play a key role in nitrogen risk management as outlined in the 2011 US nitrogen assessment (USEPA-SAB, 2011). However, the unintended effects of wetland creation (to maximize denitrification in NO_3^- enriched watersheds) on the nitrogen cycle should be carefully evaluated. For example, permanently flooded wetlands have lower N_2O/dinitrogen ratios of emissions than intermittently flooded wetlands. The ratio is also lower in warm months and warm climates.

In coastal waters, the management of algal blooms not only prevents fish kill (dead zones) but also reduces the denitrification of NO_3^- to N_2O. Because it affects the growth of phytoplankton,[17] the amount of nutrients in the euphotic zone of the oceans[18] plays a crucial role in the ocean's ability to sequester carbon (that is, in the operation of the "biological carbon pump"). However, only a few studies have focused on the importance of phytoplankton community structure to the biological carbon pump (Samarpita and Mackey, 2018).

Attention has recently been given on bringing together existing EU guidance on farming practices – which is typically separated according to environmental issue and nitrogen form – so as to foster air, water and climate co-benefits.[19] For example, over the period 2000-08 the regulation of farm practices under EU Nitrates Directive not only reduced by 16% the leaching and runoff of agricultural NO_3^-, but also the agricultural emissions of NH_3, N_2O and NO_x by 3%, 6% and 9% respectively (Velthof et al. 2014). United Kingdom agricultural GHG inventory evaluates the interactions between the various nitrogen forms when taking measurements, i.e. simultaneously measuring N_2O, NH_3, NO_3^- leaching, and nitrogen offtakes in crops. This has allowed to more fully understanding the impacts of farm management on the nitrogen cycle and more fully account for these interactions in the GHG inventory. Chapter 4. discusses in detail the unintended consequences on the nitrogen cycle of conservation practices in US agriculture.

Notes

[1] This type of analysis deserves to be expanded and deepened in the future given the importance of enhancing policy coherence.

[2] Called Modular Applied GeNeral Equilibrium Tool or MAGNET.

[3] Nitrogen fertiliser production is energy intensive (Haber-Bosch process) and production costs are highly dependent on energy prices.

[4] The 4R concept was developed by The Fertilizer Institute (TFI) in collaboration with the International Plant Nutrition Institute (IPNI), the International Fertiliser Association (IFA) and the Canadian Fertiliser Institute (IFC) as part of the nutrient stewardship initiative (www.nutrientstewardship.com/). The concept is to use the right source of fertiliser (adapt the fertiliser composition to the needs of the crop) at the right rate (quantity corresponding to the needs of the crop), at the right time (when crops need it) and in the right place (where crops can use it).

[5] See www3.epa.gov/ttnchie1/ap42/ch14/final/c14s01.pdf.

[6] As has been piloted in some countries, such as Sweden.

[7] Leaves of all higher plants have special microscopic pores on their surface, called stomata, which are important for exchange of water vapor, CO_2, and oxygen.

[8] Authored by Paul Drummond, Paul Ekins and Paolo Agnolucci of University College London (UCL) Institute for Sustainable Resources.

[9] 'Satisficing' refers to the tendency of individuals and organisations to take decisions based on habit, routines and innate assumptions (Grubb, 2014).

[10] Pollution swapping can be defined as the increase in one pollutant as a result of a measure introduced to reduce a different pollutant (Stevens and Quinton, 2009).

[11] Risk-risk trade-offs occur when interventions to reduce one nitrogen risk can increase other nitrogen risks.

[12] Assessing the effectiveness of best management practices is ultimately assessing the cost-effectiveness of instruments to incentivise them.

[13] "deNOx system" means an exhaust after-treatment system designed to reduce NO_x emissions (e.g. passive and active lean NO_x catalysts, NO_x adsorbers and SCR systems).

[14] N_2O can also be formed during conventional secondary biological treatment, which relies on aerobic processes using bacteria to remove soluble biodegradable organic matter.

[15] Changed process conditions can be caused by changes in environmental conditions (flows and loads variation for instance) and/or at the transition between anoxic and aerobic zones.

[16] The variable ratio between denitrification and anammox observed in the ocean is attributed to localised variations in organic matter quality and quantity (Babbin et al., 2014).

[17] Phytoplankton are a highly diverse group of microscopic photosynthesising microalgae and cyanobacteria.

[18] The euphotic zone is the layer closer to the surface that receives enough light for photosynthesis to occur.

[19] Joint European Commission-UNECE Workshop "Towards Joined-up Nitrogen Guidance for Air, Water and Climate Co-benefits", Brussels, 11-12 October 2016.

References

Babbin, A.R. et al. (2014), "Organic Matter Stoichiometry, Flux, and Oxygen Control Nitrogen Loss in the Ocean", *Science*, 344(6182).

Bartelings, H. et al. (2016), "Estimating the Impact of Fertiliser Support Policies: A CGE Approach", paper presented at the 19th Annual Conference on Global Economic Analysis, Washington DC, USA, June 15-17, 2016, Global Trade Analysis Project (GTAP), Department of Agricultural Economics, Purdue University, gtap.agecon.purdue.edu/resources/download/8287.pdf.

Bouwman, A.F. et al. (2002), « Modeling Global Annual N_2O and NO Emissions from Fertilized Fields", *Global Biochemical Cycles*, 16(4).

Brentrup, F. and Ch. Pallière (2009), "Energy Efficiency and Greenhouse Gas Emissions in European Nitrogen Fertiliser Production and Use", Fertilizers Europe, www.fertilizerseurope.com/fileadmin/user_upload/publications/agriculture_publications/Energy_Efficiency__V9.pdf.

Burkhardt, J and S.Pariyar (2014), "Particulate Pollutants are Capable to 'Degrade' Epicuticular Waxes and to Decrease the Drought Tolerance of Scots Pine (Pinus sylvestris L.)", *Environ Pollut.*, 184.

Danish Economic Councils (2018), "Economy and Environment, 2018", Summary and Recommendations, De Økonomiske Råd.

de Vries, W. et al. (2017), « Modelling Long-term Impacts of Changes in Climate, Nitrogen Deposition and Ozone Exposure on Carbon Sequestration of European Forest Ecosystems", *Sci Total Environ*, 605-606.

Drummond, P. et al. (2015), "Policy Instruments to Manage the Unwanted Release of Nitrogen into Ecosystems – Effectiveness, Cost-Efficiency and Feasibility", paper presented to the Working Party on Biodiversity, Water and Ecosystems at its meeting on 19-20 February 2015, ENV/EPOC/WPBWE(2015)8.

Duval, R. (2008), *A Taxonomy of Instruments to Reduce Greenhouse Gas Emissions and Their Interactions*, OECD Economics Department Working Paper N° 636.

Foley, J. and P. Lant (2009), "Direct Methane and Nitrous Oxide Emissions from Full-scale Wastewater Treatment Systems", Occasional Paper N°24, Water Service Association of Australia, Melbourne.

Görlach, B (2013), "What Constitutes an Optimal Policy Mix? Defining the Concept of Optimality, including Political and Legal Framework Conditions", CECILIA2050 WP1 Deliverable 1.1., Ecologic Institute, Berlin.

Grantz, D.A. et al. (2018), "Ambient Aerosol Increases Minimum Leaf Conductance and Alters the Aperture-Flux Relationship as Stomata Respond to Vapor Pressure Deficit (VPD)", *New Phytol.*, 30 March (Epub ahead of print).

Grubb, M. (2014), *Planetary Economics: Energy, Climate Change and the Three Domains of Sustainable Development*, Routledge, London.

Hansen, A. T. et al. (2016), "Do Wetlands Enhance Downstream Denitrification in Agricultural Landscapes?", *Ecosphere*, 7(10).

Jacobsen, B. H. et al. (2017), "Implementing the Water Framework Directive in Denmark – Lessons on Agricultural Measures from a Legal and Regulatory Perspective", *Land Use Policy*, 67.

Kampschreuer, M. J. et al. (2009), "Nitrous Oxide Emission during Wastewater Treatment", *Water Research*, 41 (17).

Karp, L. (2005), "Nonpoint Source Pollution Taxes and Excessive Tax Burden", *Environmental and Resource Economics*, 31.

Lawson, C. E. et al. (2017), "Metabolic Network Analysis Reveals Microbial Community Interactions in Anammox Granules", *Nature Communications*, doi.org/10.1038/ncomms15416.

Matthews, L. (2010), "Upstream, Downstream: The Importance of Psychological Framing for Carbon Emission Reduction Policies", *Climate Policy*, 10.

OECD (2018a), *Agricultural Policy Monitoring and Evaluation 2018*, OECD Publishing, Paris, https://doi.org/10.1787/agr_pol-2018-en.

OECD (2018b), "Evaluating the Environmental Impact of Agricultural Policies", paper presented to the Joint Working Party on Agriculture and the Environment at its meeting on 4-5 April 2018, COM/TAD/CA/ENV/EPOC(2017)14/REV1.

OECD (2013), *Water Security for Better Lives*, OECD Publishing, Paris, doi.org/10.1787/9789264202405-en.

OECD (2008), "An OECD Framework for Effective and Efficient Environmental Policies", paper prepared for the Meeting of the Environment Policy Committee (EPOC) at Ministerial Level, 28-29 April 2008, www.oecd.org/env/tools-evaluation/41644480.pdf.

Omonode, R. A. et al. (2017), « Achieving Lower Nitrogen Balance and Higher Nitrogen Recovery Efficiency Reduces Nitrous Oxide Emissions in North America's Maize Cropping Systems", *Frontiers in Plant Science*, 8(1080).

Samarpita, B. and K. R. M. Mackey (2018), "Phytoplankton as Key Mediators of the Biological Carbon Pump: Their Responses to a Changing Climate", *Sustainability*, 10(869).

Schomers, S. and B. Matzdorf (2013), "Payments for Ecosystem Services: A Review and Comparison of Developing and Industrialised Countries", *Ecosystem Services*, 6.

Stevens, C.J. and J. N. Quinton (2009), "Policy Implications of Pollution Swapping", *Physics and Chemistry of the Earth*, 34 (8-9), doi.org/10.1016/j.pce.2008.01.001.

Suarez-Bertoa, R. et al. (2016), "On-road Measurement of NH_3 and N_2O Emissions from a Euro V Heavy-duty Vehicle", *Atmospheric Environment*, 139, August 2016, doi.org/10.1016/j.atmosenv.2016.04.035.

USEPA-SAB (2011), *Reactive Nitrogen in the United States: An Analysis of Inputs, Flows, Consequences and Management Options*, U.S. Environmental Protection Agency's Science Advisory Board, EPA-SAB-11-013, USEPA, Washington D.C., yosemite.epa.gov/sab/sabproduct.nsf/WebBOARD/INCFullReport/$File/Final%20INC%20Report_8_19_11(without%20signatures).pdf.

Velthof, G.L. et al. (2014), "The Impact of the Nitrates Directive on Nitrogen Emissions from Agriculture in the EU-27 during 2000-2008", *Science of the Total Environment*, 468-469.

Von Lampe, M., et al. (2014), "Fertiliser and Biofuel Policies in the Global Agricultural Supply Chain: Implications for Agricultural Markets and Farm Incomes", OECD Food, Agriculture and Fisheries Papers, N°69, OECD Publishing, Paris, doi.org/10.1787/5jxsr7tt3qf4-en.

Chapter 6. An assessment of the effectiveness, efficiency and feasibility of nitrogen policy instruments

This chapter evaluates, generically, the pros and cons of different policy instruments for nitrogen management, and their combinations, with respect to the criteria of effectiveness, cost-efficiency and feasibility. It provides examples of evaluation of effectiveness, cost-efficiency and feasibility for a number of instruments implemented in Australia, France, Japan, Sweden and the United States.

6.1 Key findings

Drummond et al., 2016 assesses the advantages and disadvantages, in the generic sense, of the seven categories of instruments defined in Chapter 5. and their groupings. The evaluation leads to the following conclusions.[1]

First, instruments should be 'impact'-based, and applied as close to the point of emission as possible to maximise effectiveness and cost-efficiency. However, this is often not technically or administratively feasible (on some mobile or non-point pollution sources, for example). Similarly, upstream application of a pricing instrument, for example, may not be feasible for some pollutants, such as nitrogen oxides (NO_x) and nitrous oxide (N_2O) from combustion (the magnitude of which is dependent on the combustion technology, rather than a function simply of process inputs).

Second, pricing and direct environmental regulation are often most effective in reducing pollution, when subject to credible monitoring and enforcement. Whilst pricing instruments are likely to be the most cost-efficient policy option to achieve a given level of abatement (and direct regulation often the least), they are relatively 'blunt' instruments, and may create pollution 'hotspots' across space, time and sectors (depending on the scope of the instrument). 'Tax aversion', or the relatively low public and political acceptability of pricing instruments (in particular, but also other 'polluter pays' instruments)[2] often leads to the introduction of exemptions, discounts and other measures (e.g. permit grandfathering), to secure support, reducing cost-efficiency. Polluter pays instruments may also lead to the Pollution Haven Effect,[3] although the evidence for the existence and strength of this phenomenon is mixed.

Third, 'beneficiary pays' instruments)[4] (e.g. public financial support), along with voluntary and informational instruments are often more politically feasible than 'polluter pays' instruments (e.g. pricing mechanisms), due to the lack of the direct cost to firms and the circumvention of issues contributing to tax aversion - although they are likely to be less effective (and in the case of financial support, in particular, less cost-efficient).

Fourth, combining 'polluter pays' with 'beneficiary pays' instruments may be more politically feasible to introduce to achieve a given level of environmental effectiveness than either employed individually. A key example is through the use of a 'feebate' instrument.[5] The risk of inducing a Pollution Haven Effect is also reduced against the use of an equivalent 'polluter-pays' instrument introduced in isolation.

Fifth, combining pricing or public financial support instruments with direct environmental regulation may be mutually beneficial for various reasons. The application of the direct regulation as a 'secondary' instrument to support a 'primary' pricing or public financial support instrument reduces vulnerability to market distortions (such as split-incentives and environmentally harmful subsidies), but also the potential for pollution hotspots. Targeted secondary direct regulation on a sub-set of participants to the primary pricing instrument reduces the administrative complexity required if attempting to tackle hotspots through complex compliance rules (e.g. differentiated tax rates, access rules, permit multipliers, etc.) within the primary instrument itself. The cost-efficiency (and potentially effectiveness) of such combination is thus increased against the use of the primary instrument alone. The application of the pricing or public financial support instruments as secondary instruments to a primary direct regulation may increase cost-efficiency for a given level of abatement against the use of a direct regulation alone. Public and political acceptability may be increased in both instances.

Sixth, voluntary and information instruments may require the lowest administrative capacity (and produce the least transaction costs), although transaction costs and political acceptability for the introduction of an effective voluntary negotiated agreement may be prohibitive. Voluntary instruments (in particular) may also experience the greatest risk of regulatory capture.[6] However, the use of information instruments is likely to increase the effectiveness, cost-efficiency or feasibility (or a combination of these assessment criteria) when combined with all other instrument categories.

A summary of results is provided in Table 6.1. Each instrument category displays very different characteristics against each of the three assessment criteria, with further differences induced depending on: (i) the specific design of the instrument concerned; (ii) the type of pollution the instrument is applied to; (iii) the source of the pollution in question (e.g. stationary, mobile or non-point); and, (iv) the institutional context within which it operates. Trade-offs between each of the three criteria are therefore unavoidable in a second-best world, characterised by market failures, uncertainties and practical constraints.

As such, in order to maximise effectiveness and cost-efficiency within 'feasible' constraints, an instrument mix is often required (OECD, 2007). Table 6.2 summarises the key pros and cons of each instrument pairing against the three assessment criteria. It is assumed that each instrument in the combination presented is present at the same point in time (although not necessarily introduced at the same time).

Table 6.1. Effectiveness, efficiency and feasibility of policy instruments

	Policy effectiveness	Policy cost-efficiency	Policy feasibility
Environmentally related taxes and charges	Effective in reducing pollution, but to varied levels, and difficult to estimate ex ante (particularly upstream taxation) Does not address hotspots Pollution haven effect possible	Statically cost-efficient if applied equally across emitters, but often policies allow exemptions and discounts Dynamic cost-efficiency generally high (continuous liability), but depends on tax rate and policy certainty	May be applied upstream, at point of emission or further downstream Administrative feasibility high, as used in all countries for different purposes. Upstream more so, as fewer actors (also less tax avoidance), but impractical for NO_x and N_2O from combustion. Downstream more efficient, but difficult to monitor (particularly non-point source) World Trade Organisation (WTO) trade rules may place limits on design; other international/supranational conventions may prevent some instruments (e.g. taxation of aviation fuel), along with pricing of previously unpriced emissions Significant public aversion to taxation Essential to address competitiveness and distributional concerns often essential to address (e.g. with exemptions/discounts) to gain acceptance, reducing effectiveness/efficiency; environmental taxation reform principles may avoid this
Tradable permit systems (TPS)	Effectiveness known ex ante, if well designed and enforced; have been effective in reducing (particularly air) pollution Can't address hotspots Pollution haven effect possible	Statically cost-efficient if applied equally across emitters, but often provide permits for free, producing windfall profits, producing potential perverse incentives; market power often increases costs Dynamic efficiency mixed – depends on level of cap/baseline and change over time, exogenous developments, future permit allocation methodologies	May be applied upstream, at point of emission or further downstream Administrative requirements probably higher than a tax, due to additional complexity; may be prohibitive for all but large stationary installations Generally, high-income countries have experience with these instruments, low-income countries do not; potential administrative capacity/acceptance issues WTO and legal issues (described under environmentally related taxes and charges) Essential to address competitiveness and distributional concerns, with free permit allocation and exemptions common features (reducing effectiveness/efficiency)

	Policy effectiveness	Policy cost-efficiency	Policy feasibility
Direct environmental regulation (DER)	Effective in achieving direct objectives, but environmental effectiveness depends on definition of this objective Can overcome other market failures (e.g. split incentives) that prevent market-based instruments achieving higher efficacy Long-term impact (if instrument is removed) more likely than with other instruments Can address hotspots Pollution haven effect possible Rebound effects possible	Statically and dynamically cost-inefficient (particularly 'outcome'-based technology/practice prescriptive or prohibitive regulation) Issues reduced if regulation is 'impact'-based, with clearly increasing stringency over time	Can address broad spectrum of pollution from point and non-point sources May require significant administrative capacity WTO trade rules/other supranational rules may place limitations on use Higher potential for regulatory capture with pricing instruments (particularly in low-income countries) Potential competitiveness and distributional concerns as with pricing instruments Low flexibility
Public financial support (PFS)	Effectiveness highly dependent on context factors, but with significant potential Clear eligibility criteria can prevent hotspots Rebound effects possible	'Beneficiary' rather than 'polluter' pays More cost-efficient than direct regulation, but less than pricing instruments – although depends against what the public financial support is provided Design should avoid deadweight costs as far as possible Dynamic cost-efficiency depends on focus of instrument, and whether support levels are reactive over time	Public resources must be available to fund the instrument, with consequential political challenges Financial support instruments often preferred – avoids direct cost burden and avoids tax aversion issues For lending instruments, authority must have capacity to bear debt WTO/supranational rules (including state aid) may place restrictions on use Potentially high flexibility
Payments for ecosystem services (PES)	Effectiveness often positive, but relatively limited (although relatively little robust assessment of PES instruments have been made) Pollution haven effect possible	Effectively a financial support instrument – similar cost-efficiency implications User-funded rather than government-funded systems likely to be more cost-efficient (particularly at small scale)	Well-defined access criteria, baseline and performance metrics required Difficult to measure output of services – proxies may need to be used Other objectives, such as poverty alleviation, may be required for Government funded systems, but reduce environmental effectiveness Consistent funding stream required Commonly used in low-income countries, where administrative capacity is weaker Potentially high flexibility

	Policy effectiveness	Policy cost-efficiency	Policy feasibility
Information measures	Often difficult to assess effectiveness individually, as usually introduced or combined with other instruments Can improve effectiveness of other instruments Effective in revealing low- or negative-cost abatement opportunities	Static and Dynamic cost-efficiency depends on type of information instrument considered and scope of application (e.g. sectors with few non-informational barriers) Generally low, and potentially negative cost to industry and consumers Can increase efficiency of other instruments	Administrative requirements depend on instrument – e.g. Pollutant Release and Transfer Registration (PRTR) systems require significantly more capacity than information campaigns WTO rules (treatment of 'like' products) Potentially high flexibility (depending on instrument and design)
Voluntary schemes	Effectiveness often low, although depends on specific instrument Various conditions required for voluntary agreement effectiveness, including threat of mandatory alternative, and limited scope of application	Voluntary agreements unlikely to meet three conditions for static or dynamic cost-efficiency (full sector coverage, full information disclosure and lowest-cost burden distribution between firms)	Voluntary instruments may have greatly reduced administrative burden (although transaction costs for an effective negotiated agreement may be prohibitive), and may be legally binding or non-binding High potential for collusion and regulatory capture

Source: Drummond et al. (2015a).

Table 6.2. Effectiveness, efficiency and feasibility of policy instrument combinations

	Environmentally related taxes and charges	Tradable permit systems (TPS)	Direct environmental regulation (DER)	Public financial support (PFS)	Payments for ecosystem services (PES)	Information measures
Tradable permit systems (TPS)	Aggregate abatement unaffected 'Hybrid' instrument may reduce static, but improve dynamic cost-efficiency May improve flexibility and ability to deal with uncertainty- can reduce price/cost uncertainty Can tackle hotspots (potentially increasing efficiency) Political feasibility may be increased					
Direct environmental regulation (DER)	May increase effectiveness (overcome market failures, and reduce rebound effect) Can tackle hotspots DER can increase static cost-efficiency if overcomes market failures DER can reduce uncertainty of abatement under a tax/charge May reduce administrative complexity May increase political feasibility	Aggregate abatement unaffected DER can prevent against hotspots under a TPS and increase cost-efficiency TPS can increase flexibility for compliance with a DER, and increase cost-efficiency - also increased if overcomes market failures (e.g. split incentives). Both improve public/political feasibility Regardless of the primary instrument, ability to deal with uncertainties likely increased				

6. AN ASSESSMENT OF THE EFFECTIVENESS, EFFICIENCY AND FEASIBILITY OF NITROGEN...

	Environmentally related taxes and charges	Tradable permit systems (TPS)	Direct environmental regulation (DER)	Public financial support (PFS)	Payments for ecosystem services (PES)	Information measures
Public financial support (PFS)	May overcome market failures (e.g. split incentives) to increase effectiveness and cost-efficiency. Likely less cost-efficiency than 'full-tax' option, reduces average total costs for firms (encouraging entry) Can tackle hotspots (more feasible than full 'polluter pays' approach, but possibly less effective), and may reduce risk of Pollution Haven Effect against use of tax/charge alone Can tackle innovation failures, increase dynamic cost-efficiency Increased feasibility, possibly revenue neutral. May reduce flexibility	Aggregate abatement unaffected under TPS Can tackle hotspots, and may reduce likelihood of pollution haven effect against use of TPS alone Cost-efficiency reduced with PFS under TPS, unless tackling market failures (e.g. split incentives, innovation), pollution hotspots or innovation failures PFS can be used to allocate permits, may reduce windfall rents Increased feasibility with use of 'beneficiary-pays', possibly revenue neutral and lower average cost to firms than tax/charge alone (but encourages entry). May reduce flexibility	May be more effective and politically feasible than each introduced individually PFS may increase overall cost-efficiency against direct regulation alone (also through tackling innovation failures)– although a cost efficient design overall difficult to achieve Direct regulation provides increased certainty against PFS instrument alone, and may increase cost-efficiency (particularly if targeting hotspots) May allow for less complex instrument designs			
Payments for environmental services (PES)	As a PES is a PFS instrument, dynamics generally as above However, likely to be less effective and cost efficient due to indirect and uncertain nature of a PES	As a PES is a PFS instrument, dynamics generally as above	As a PES is a PFS instrument, dynamics generally as above	As a PES is a PFS instrument, not examined in this paper		

	Environmentally related taxes and charges	Tradable permit systems (TPS)	Direct environmental regulation (DER)	Public financial support (PFS)	Payments for ecosystem services (PES)	Information measures
Information measures	Environmental effectiveness mutually re-enforced Pollution hotspots may remain Cost-efficiency increased, by overcoming information failures; optimal tax/charge level reduced. May allow for improved initial design May increase public and political acceptability as reduced cost (at given level of abatement)	Aggregate abatement unaffected By reducing information failures, cost-efficiency increased. May allow for improved initial TPS design Due to this, political feasibility increased and the potential for the pollution haven effect may be reduced against a TPS alone (but short-term dynamic cost-efficiency may decrease) Pollution hotspots may remain	May act together to 'push' and 'pull' actors towards abatement/lower pollution intensity Information may improve cost-efficiency of direct regulation, and initial design. As such, risk of pollution haven effect may be reduced and political feasibility of direct regulation increased Pollution hotspots may remain	Environmental effectiveness mutually re-enforced (but less so than with a tax/charge) Cost-efficiency increased, by overcoming information failures; optimal support level reduced May allow for improved PFS design Pollution hotspots may remain Unlikely to induce Pollution Haven Effect (and may have opposite influence) Highly acceptable to recipients, as no 'polluter-pays' element	As a PES is a PFS instrument, dynamics generally as with PFS/information measure combination	
Voluntary schemes (VS)	Less effective and cost efficient than tax/charge alone, but more politically feasible 'Awareness effect' from VS may improve effectiveness of tax/charge Hotspots' may be tackled, but unlikely to be effective Administrative burden of effective combination likely to be high	VS under a TPS will not affect aggregate abatement, but may tackle hotspots (although perhaps ineffectively), and may encourage an increase in regulatory reach, increasing cost-efficiency. Otherwise, cost-efficiency reduced. If a VS used to distribute free permits under a TPS, short-term abatement uncertainty may be reduced If TPS used to achieve objectives of a VS, effectiveness and cost-efficiency is highest against other forms of VS, but low feasibility	Effectiveness likely to be reduced against the use of a direct environmental regulation alone, but cost-efficiency and political feasibility may increase May tackle hotspots (but VS as secondary to DER may be ineffective) Cost-efficiency unclear, and depends on design of each component Administrative burden may be significantly increased, if effective and efficiency voluntary agreement is in place Flexibility and ability to deal with uncertainty may be increased against direct environmental regulation alone	PFS may encourage additional participation in a VS. VS unlikely to increase effectiveness of PFS Cost-efficiency against use of either instrument alone likely reduced PFS may overcome innovation failure, may increase long-term cost-efficiency of VS (but such a combination has low feasibility) Administrative burden of an effective, cost efficient instrument likely to be high	As a PES is a PFS instrument, dynamics generally as with voluntary scheme/PFS combination	Provision of information (both to polluters, and disclosure and reporting requirements) likely to increase effectiveness of VS Information likely to increase cost-efficiency of a VS Pollution hotspots may remain Feasibility of a VS increased with provision of information to polluters, possibly reduced with reporting/disclosure requirements

Source: Drummond et al. (2015b)

6.2 Case studies of policy instruments

Such 'generic' assessment of each instrument category (or groupings thereof) against each of the three assessment criteria based on key design options may be broadly applicable to any type of pollutant, and applied to tackle the same pollutant, produced or released via the same pathway(s). In the case of nitrogen pollution, the analysis may be broadly applied to the release of nitrogen into the atmosphere from the combustion of fossil fuels, or into water bodies from the application of fertiliser, for example. Case studies of where such instruments or their combinations have been applied to reduce anthropogenic disturbance of the nitrogen cycle are presented below. However, discussion of the complex interaction between different pathways of nitrogen pollution, and appropriate policy mixes that may be required to most appropriately tackle this, is beyond the scope of this Chapter.

6.2.1 The Swedish refund emission payment for nitrogen oxides (NO$_x$): a combination of environmentally related tax and public financial support (PFS)

Since the 1980s, soil acidification has been a significant political issue in Sweden, a country, due to its underlying geology, particularly susceptible to acid deposition via acid rain, damaging both aquatic and terrestrial ecosystems. In order to partially address this, in 1992 the Swedish government introduced a tax on measured NO$_x$ emissions from stationary large combustion plants producing at least 50 GWh of useful energy annually (impacting around 200 plants), with exemptions for activities for which costs would have been excessive, such as the cement and lime industry, mining, refineries, blast furnaces, the glass and insulation material industry, wood board production and the production of biofuels (OECD, 2013). The tax was part of a wider strategy to reduce Swedish NO$_x$ emissions by 30% between 1980 and 1995 (Höglund-Isaksson and Sterner, 2009). The tax base was later increased with a threshold for entry reduced twice; the tax is currently applicable to plants with an annual useful output of at least 25 GWh (around 400 plants) (OECD, 2013). The tax rate was initially set at SEK 40 (USD 5.5)/kg NO$_x$ emitted, an exceptionally high rate compared to other NO$_x$ pricing systems in the OECD area (e.g. 200 times the rate levied in France) (Ecotec, 2001). In 2008, the rate was increased to SEK 50 (USD 6.4)/kg NO$_x$. All revenue from the tax is recycled back to participants[1] in proportion to their production of useful energy, meaning that firms with low (high) NO$_x$- intensity per unit of energy produced become net beneficiaries (losers) (OECD, 2013). As such, the instrument is often referred to in the literature as a 'Refund Emission Payment'.

Instrument effectiveness. NO$_x$ emissions from obligated installations fell by over 50% between 1980 and 1997 (coupled with rapidly increasing energy output)[2] (OECD, 2013), with the 1992 NO$_x$ tax likely responsible for at least two-thirds of this (with other local regulation also having some influence), indicating high effectiveness (Ecotec, 2001). However, more recent studies suggest that total emissions from obligated entities have decreased only marginally between 1992 and 2013, with total emissions from the net 'losers' decreasing by around 30%, largely offset by increased total emissions from net 'winners'. However, emission intensity (NO$_x$/KWh) have more than halved in the same timeframe, driven particularly by the net 'winners' (Naturvårdsverket, 2014).

Instrument cost-efficiency. An environmental tax on NO$_x$ applied downstream requires continuous emission monitoring, which is infeasible for mobile sources, and expensive for small installations (OECD, 2013), with the result that the instrument has only been applied to relatively large stationary combustion installations. Whilst this reduces overall efficiency, the market-based pricing approach provides static cost-efficiency across participating polluters with highly heterogeneous marginal abatement costs (OECD, 2013). However, basing the Refund Emission Payment on the proportion of energy produced is cost-inefficient from the perspective of NO$_x$ abatement, and instead implies a subsidy to energy generation, incentivising excessive production and investment in some facilities (e.g. waste incineration plants). Dynamic cost-efficiency is also relatively high, with the continued incentive to abate delivered by the initial taxation, and to some degree, the Refund Emission Payment. Innovation as a result of the tax has been rapid, with the production and diffusion of new technologies, alongside the adoption of existing best-practice technologies, found to be 'very important' in delivering the NO$_x$ abatement achieved (Höglund-Isaksson and Sterner, 2009; Sterner and Turnheim, 2008). The Refund Emission Payment may have inhibited further diffusion of new innovations, as an individual firm is incentivised to retain the innovation to reduce NO$_x$ emissions as far as possible relative to other regulated installations, in order to obtain a higher rebate level (Höglund-Isaksson and Sterner, 2009). However, recent analysis questions these findings, and argues that inhibited innovation and diffusion of technology may have occurred in the absence of a refund mechanism (Bonilla et al., 2014, SOU, 2017). In 2011, total revenues amounted to SEK 794 million (USD 101 million) (OECD, 2013).

Instrument feasibility. The tax had broad public and political support when proposed and introduced. The impact of NO$_x$ on the environment was visible and widely understood, and the tax was proposed by an all-party parliamentary commission that also included representatives from all relevant ministries and civil society. The high tax rate introduced was made possible by these factors, but also by the innovative recycling mechanism, along with the exemption of industries with potential competitiveness issues (OECD, 2013).

6.2.2 Selected instrument combinations of relevance to nitrogen pollution

Environmentally related taxes and charges and tradable permit systems (TPS) (combination cost-efficiency)

Both taxes and charges and TPS may lead to the concentration of pollutants geographically, sectorally and temporally (i.e. the creation of a pollution 'hotspot')[3] (Drummond et al., 2015b). Whilst this does not matter for pollutants such a greenhouse gases (GHGs), for some pollutants, the specific conditions of their release may have substantial impact on their marginal damages. For example, a unit of NO$_x$ in the atmosphere will have substantially higher marginal costs (via impacts on human health) in a dense, highly populated urban area than in a remote, rural location. The application of an additional tax or charge to firms participating in a tradable permit system, and which are located in an air/watershed where the emission of a given pollutant creates marginal damages higher than for participants in other air/watersheds, may tackle this issue.

Environmentally related taxes and charges and direct environmental regulation (DER) (combination effectiveness)

A substantial share of environmentally related taxes and charges are used in combination with at least one regulatory instrument (OECD, 2006). DER is often able to overcome certain market failures, such as principal-agent problems, information and other market failures (including environmentally harmful subsidies) that inhibit the effectiveness of environmentally related taxes and charges (and other market-based mechanisms). Conversely, such taxes and charges are able to reduce the 'rebound' effect[4] that direct regulation may induce. Such interactions generally hold true whether each instrument directly targets the same group, or whether the instruments are applied to different actors but aim to tackle the same pollution source (e.g. upstream pricing of fossil fuels with downstream energy efficiency requirements). As such, if well designed, implementing such instruments in combination may have positive, mutually re-enforcing effects on environmental effectiveness, cost-efficiency and feasibility.

Direct regulation combined with a pricing instrument may also reduce the occurrence of pollution hotspots. For example, whilst pricing instruments influence total pollution, direct regulation may influence the specific characteristics (e.g. location and timing) of its release. For example, an 'impact-based'[5] regulation may set localised ambient quality standards, whilst an 'outcome-based'[6] approach may ban or require the use of certain technologies or specified activities, to prevent or reduce location- or time-specific environmental harm that a broad pricing instrument would not have incentivised against.[7] Such instrument combinations have been widely and successfully deployed for this purpose (OECD, 2006; Bennear and Stavins, 2007). However, such a combination may potentially induce a Pollution Haven Effect without sufficient compensatory measures.

Environmentally related taxes and charges and public financial support (PFS) (combination effectiveness)

From a theoretical standpoint, the marginal incentives for pollution abatement provided by a tax or charge instrument (embodying the 'polluter pays' principle) and a PFS instrument (embodying the 'beneficiary pays' principle) are identical, if the tax/charge and support rates are set at the same level, and the externality is targeted directly (i.e. impact-based) (Baumol and Oates, 1988). However, even a well-designed PFS instrument would likely be less environmentally effective than an equivalent tax or charge. This is because whilst a 'polluter pays' instrument such as a tax acts to increase average total costs to the firm, PFS instruments act to decrease them. This allows firms (e.g. older, less efficient and more polluting firms) to maintain operation that may have otherwise been forced out of the market by a taxation instrument. Additionally, PFS does not allow the market to communicate the true cost of the product to the consumer, maintaining sub-optimal demand (too high, in the case of a negative externality) (Kolstad, 2011). Such issues maintain pollution levels above that expected from an equivalent taxation instrument.

As such, it would appear nonsensical to deploy both instruments to reduce a given pollutant from a given source, when a single instrument (particularly a tax or charge) could achieve the same objective. However, combining such instruments in a well-designed approach may increase the likelihood of achieving the desired abatement that feasibility constraints prevent a single instrument from achieving.

A common approach is through the use of 'feebate' instruments (or 'bonus-malus'), which levy a tax or charge ('fee') on activities or products below a given environmental threshold performance or limit, and financial support to activities or products that exceed this performance level ('rebate'). Figure 6.1 illustrates the design of a feebate instrument. In this example, the threshold (or 'pivot point') above which a 'fee' is due and below which a 'rebate' is given is 30 units of pollution. The Y-axis illustrates absolute values, meaning the gradient of the line indicates the marginal abatement costs and benefits (i.e. the rate of the tax and rebate due). Also illustrated in Figure 6.1 is the design of a similar tax/charge-only approach. As is clear, the gradient of the two slopes is the same, indicating equivalent marginal incentives, with the reasoning behind the differences in average total costs clearly demonstrated.

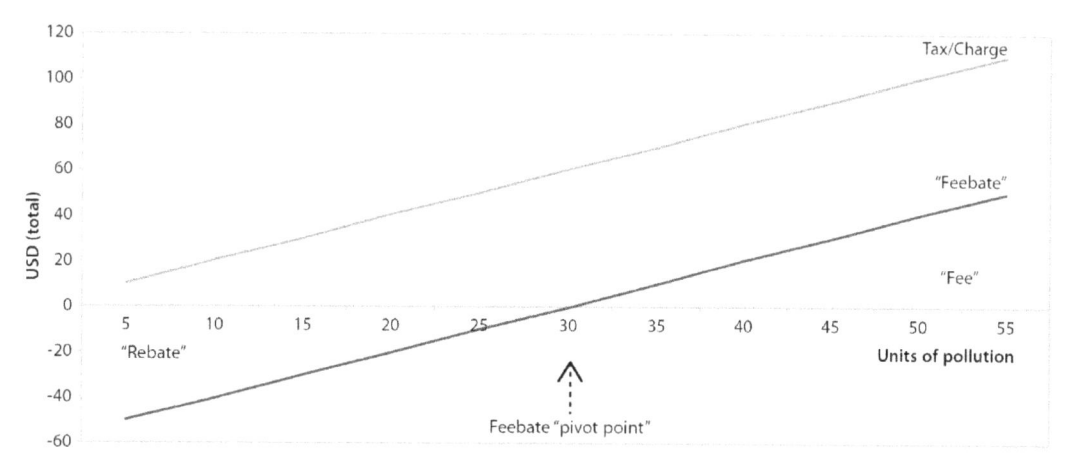

Figure 6.1. Design of a 'feebate' instrument against a tax/charge

Source: Drummond et al. (2015b).

Whilst feebates may be applied to a wide range of pollutants from different sources (using a variety of specific designs), they are most commonly investigated for and applied to passenger car registration duties (against carbon dioxide (CO_2) intensities). In this function (and more broadly), the evidence suggests well-designed examples are generally successful in improving environmental performance (Greene et al., 2005; Johnson, 2006; de Haan et al., 2009). However, this depends on the specific design of the instrument, including the pivot point and the levels of fees and rebates set (D'Haultfoeuille et al., 2014).[8]

Environmentally related taxes and charges and payments for ecosystem services (PES) (combination effectiveness)

As a PES instrument may be considered a specific category of PFS instrument, combining environmental taxes or a charge with a PES instrument produces dynamics similar to those produced by a broader tax/charge and PFS combination, discussed in the section above. This applies both to 'government-funded'[9] and 'user funded'[10] PES approaches (Drummond et al., 2015b).

However, some key specifics may be highlighted. Although relatively few rigorous ex post analyses of PES instruments have been undertaken (Drummond et al., 2015a), those that are available indicate a generally positive, but limited effectiveness (Pattanayak et al., 2010). Despite this, combining a tax or charge (or tradable permit system) with a PES instrument is likely to be less environmentally effective than a combination with other PFS approaches (particularly impact-based), both in achieving a targeted level of abatement, and in tackling pollution hotspots, in which, as with a tax/charge and PFS combination, the tax/charge is primary, with the PFS instrument secondary.

The key reason for this is the difficulty inherent in defining and directly targeting the ecosystem service in question. This often requires a detailed understanding of causal pathways (recognising spatial extent and distribution) that are yet either not fully understood, or impractical to monitor (Tomich et al., 2004). As such, proxies or indicators are commonly used (i.e. an 'outcome-based' approach). This links to the second issue; the difficulty in accurately defining the baseline to ensure 'additionality' of supported actions, and providing a suitable level of payment to encourage uptake (e.g. set at least at the differential between the private economic value gained from the presence of the ecosystem service (e.g. forest conservation), and the most profitable use of land to the owner (e.g. conversion to pasture). In other words, the opportunity cost of maintaining the ecosystem service in question (Drummond et al., 2015a)). As would be expected, preventing excessive support is equally as challenging, with consequences for cost-efficiency.

Environmentally related taxes and charges (combination cost-efficiency)

Whilst this Chapter discusses combinations of instrument between different instrument categories, it is often possible to effectively combine instruments from within the same category, but of different design. For example, in many OECD countries wastewater charges are applied to households, with a pollution tax levied further downstream on releases to water bodies from wastewater treatment plants (WWTPs). In a 'first-best' world, a cost-efficient approach would be to apply an impact-based tax at the point of wastewater release from the WWTPs, with the cost of the tax or abatement measures installed (with marginal abatement costs at or lower than the tax level) passed upstream to users (e.g. households) based on their individual pollutant contribution (plus other associated operating costs). These users would then also hold an incentive to reduce the level of wastewater (and concentration of pollutants in the wastewater) to the economically optimal level. However, determining the individual pollution contribution of users would in most cases by administratively and cost prohibitive, preventing the effective transmission of this price signal. Instead, a charge is levied on users often based on wastewater volume rather than pollutant load (and actually often based on metered water use rather than wastewater discharge, with the assumption of parity), with the objective of reducing overall wastewater discharge. An impact based-pollution tax, where present, then incentivises the WWTP to treat the remaining discharge to the economically optimal level (based on the level of the tax), before release into a water body.

Direct environmental regulation (DER) and public financial support (PFS) (combination cost-efficiency)

In January 2018 a political agreement was made concerning a new and more 'targeted regulation' of the agricultural emissions of nitrogen in Denmark waters.[11] One key element of the new DER is to specifically target efforts according to geographic differences. This contrasts with the previous DER scheme, where all farmers where required to reduce their emissions by the same amount, regardless of geographical variation in the nitrogen balance. The new DER involves dividing up the country into approximately 3 000 distinct areas and setting up a nitrogen emissions reduction target for each one in order to protect groundwater and surface water. Under the regulation, farmers are free to choose the most cost-effective tool for their locality.

The new DER is combined with PFS for agricultural production that reduces the emissions of nitrogen (e.g. catch crops, which are sown after harvesting in order to take up soil nitrogen). Despite a more targeted approach, such subsidy based regulation does not create a long-term incentive to relocate the production with the highest emissions to areas in Denmark where the emissions will damage the aquatic environment the least (Danish Economic Councils, 2018). This disadvantage could be avoided if the DER instead incurred geographically differentiated taxes on those agricultural activities which emit nitrogen to the aquatic environment (ibid).

Such a tax on agricultural activities that release nitrogen into the aquatic environment could relatively easily be extended to other types of agricultural emissions, such as ammonia (NH_3) or N_2O (Danish Economic Councils, 2018). Nitrogen taxes, as opposed to taxes on GHG emissions, will need to vary geographically; this should not increase the complexity of the regulation from the farmer's perspective since there would still be only one tax for each type of activity (ibid). In the long run such extended taxation would provide a clearer and more direct common regulation of different environmental effects from agricultural production (ibid).

6.2.3 The Greater Miami Watershed Trading Programme: an example of tradable permit system (TPS)

Around 40% of surface waters in the Great Miami River watershed of around 10 000 km² in Southwest Ohio, United States, were consistently failing to meet regulatory water quality standards. In response, the regional water management agency, the Miami Conservancy District (MCD), in 2004 introduced the Greater Miami Watershed Trading Programme (GMWTP) as a pilot scheme for a cost-effective approach to improving water quality, in anticipation of regulatory measures to set nutrient release limits on point-source WWTPs.

The GMWTP takes a baseline-and-credit approach, and encourages farmers to adopt voluntary 'best-practice' measures to generate Emission Reduction Credits (ERCs), that may be traded to WWTPs to allow for future regulatory compliance. Agricultural activities cover around 70% of land in the watershed, are the primary contributor to the excess nutrient release (Newburn and Woodward, 2012), and according to Kieser and Associates (2004), are able to achieve nutrient reductions at a cost around thirty times lower than WWTPs. Any agricultural operation within the watershed is eligible to implement measures to generate credits. A farmer, in co-operation with the local Soil and Water Conservation District (SWCD) authority, proposes a bid suggesting practice changes, the value requested to implement the changes, and calculates the number of credits these actions generate (1 ERC = 1 lb (0.45 kg) avoided nitrogen or phosphorus release), based on annual savings against a projected baseline of nutrient release, multiplied by the number of years the proposed actions are expected to generate savings. Phosphorus and nitrogen credits are fully fungible. Credits are then purchased from farmers (via SWCDs) by the MCD (which acts as a clearinghouse) by 'reverse' auction,[12] which may then be bought by WWTPs (although only credits generated by activities upstream from the purchasing WWTP may be acquired) (Newburn and Woodward, 2012; Shortle, 2012).

Instrument effectiveness. By mid-2014, 397 agricultural projects were contracted, generating over 1.14 million credits worth over USD 1.6 million, and producing an estimated 572 metric ton reduction in nutrient discharges to surface waters in the watershed against the counterfactual (MCD, 2014). As binding legislation on point sources has yet to be introduced, these transactions are driven by trading ratios that favour action in advance of expected legislative requirements.[13]

Instrument cost-efficiency. Trades completed have thus far produced a cost of USD 1.48/lb of nutrient abated. This is significantly less than the estimated USD 4.72/lb abated estimated for measures implemented directly by WWTPs (Kieser and Associates, 2004), suggesting significantly higher cost-efficiency than a regulatory approach applied to point sources only. However, whilst the use of a reverse auction should induce competition between suppliers and reveal reserve prices, it is likely that over successive auction rounds strategic bidding came into play, with SWCDs and farmers learning how to achieve the highest price whilst still having their bid accepted. This was likely facilitated by the lack of variation in the maximum bid price set by the MCD over successive rounds (USD 2/lb), reducing static cost-efficiency. The incentive for innovation has been low, as the credit calculation methodology relies on data only available for relatively common nutrient management practices (Newburn and Woodward, 2012).

Instrument feasibility. The cost-effective nature of the GMWTP compared to other regulatory options for achieving anticipated regulatory nutrient release limits makes the GMWTP popular amongst WWTPs. During formulation of the GMWTP, over a hundred meetings took place with a wide range of stakeholders, leading to wide acceptance and support for the instrument. The use of the SWCDs as agents also actively helped advertise and encourage the participation of individual farmers (alongside subsequent annual monitoring, the administrative cost of which is factored in to bids). This reduces both transaction costs and administrative burden. These factors combine to make the GMWTP a popular and well-functioning instrument (Newburn and Woodward, 2012).

6.2.4 Japan's automobile 'nitrogen oxides (NO_x) law': an example of direct environmental regulation (DER)

Despite various regulations placed on stationary sources of local air pollutants in Japan, levels of NO_x pollution continued to rise throughout the 1980s, as a result of increasing vehicle traffic. In response, the 1992 'Law Concerning Special Measures for Total Emission Reduction of Nitrogen Oxides from Automobiles in

Specified Areas' (NO$_x$ Law) was enacted, which required the prefectures of the large urban areas of Tokyo and Osaka to establish and implement local plans to reduce NO$_x$ emissions from vehicles, with the overarching aim of achieving national ambient nitrogen dioxide (NO$_2$) concentration standards by 2000 in these prefectures (a 27% decrease from 1990 levels). The law includes, in particular, 'vehicle type regulation', which prohibits both the use and registration of vehicles that fail to meet specified emission standards in the obligated prefectures (with specific 'retirement' dates differentiated by vehicle type). However, a 10% increase in vehicle mileage coupled with underperformance of other management measures (including preferential taxation, low-interest loans and investment in alternative fuel stations to promote the use of low-emission vehicles) produced a reduction in NO$_x$ emissions of just 3% (OECD, 2002). In 2001, as a countermeasure, particulate matter was added as a target substance, and Nagoya was added as an additional obligated prefecture. These amendments were codified as a revised 'Law Concerning Special Measures for Total Emission Reduction of NO$_x$ and particulate matter (PM) from Automobiles in Specified Areas' (NO$_x$ and PM Law).

Instrument effectiveness. NO$_x$ emissions decreased around 20% between 2000 and 2009 in the areas subject to the revised law, with reductions experienced at twice the rate experienced outside the obligated prefectures (Hasunuma et al., 2014). Iwata et al. (2014) finds that the vehicle type regulation, by banning the registration of high-polluting vehicles, and forcing the replacement of those already in circulation, was the key driver behind this result. The prevalence of asthma in obligated areas decreased significantly at -0.073%/year, with the induced NO$_x$ (and PM) decrease significantly correlated to this trend (Hasunuma et al., 2014).

Instrument cost-efficiency. Whilst there are no ex post cost estimates of the 'NO$_x$ Law', ex ante estimates suggest a total cost of up to JPY 521 billion (around USD 5 billion) (Iwata and Arimura, 2008; Iwata et al., 2014). The total social benefits of the regulation, however (including those of reduced health impacts, and spillover impacts to neighbouring, non-obligated regions), are almost certainly likely to have exceeded this cost (by at least double, according to lower-bound estimates by Arimura and Iwata, 2006). However, there is significant variation between marginal abatement costs among vehicle type (with standard passenger cars experiencing a marginal abatement cost more than double that of a small bus, for example), suggesting a significant potential for both reduced compliance cost and increased net social benefit by changing the schedule of retirement enforcement (Arimura and Iwata, 2006).

Instrument feasibility. Rapid economic growth in Japan from the mid-20th Century produced significant growth in air pollution, and consequential health impacts on the population (particularly in large urban areas). Public pressure, along with an increasing number of successful lawsuits filed by organisations representing the interests of victims of pollution-related health damage, led to a succession of anti-pollution laws from the 1960s onwards. The awareness of pollution-related health issues in the population is likely to have provided support for the NO$_x$ law, along with differentiated retirement dates and the presence of parallel instruments (such as graded vehicle taxation rates and subsidies for low-emission vehicles), although the literature is lacking on the extent to which public support was present at its introduction, and the contribution of these factors. The instrument is enforced through the Japanese Vehicle Inspection Programme; non-compliant vehicles cannot undergo this mandatory inspection if they are not in compliance with the NO_x law (Colls and Tiwary, 2010).

6.2.5 Pennsylvania's Resource Enhancement and Protection Programme: an example of public financial support (PFS)

Established in 2008, the Resource Enhancement and Protection Programme (REAP) provides tax credits to farmers, business and individuals in return for implementing pre-approved 'Best Management Practices' (BMPs) in agricultural operations in Pennsylvania that enhance farm production and protect natural resources (exceeding legal requirements). Applicants may receive between 50% and 75% of private implementation costs as Pennsylvania state tax credits (up to USD 150 000, depending on the BMP implemented, with publicly-funded costs ineligible), issued upon project completion. Non-agricultural businesses and individuals (subject to taxation by the Commonwealth of Pennsylvania)[14] are able to fund BMP measures and receive the associated tax credits, which are valid for fifteen years, and available for transfer to other tax payers. Credits are repayable if the practice is not continued for the pre-agreed lifespan.

Instrument effectiveness. Between 2008 and the end of 2011, over 950 agricultural operations had received tax credits for introducing at least one BMP measure, reducing estimated emissions of nitrogen to water bodies by over 5 700 metric tons (along with a reduction of phosphorus and sediment by around 430 tonnes each) (PSCC, 2011).

Instrument cost-efficiency. Over this period, REAP participants claimed tax credits worth around USD 39 million (PSCC, 2011), allowing them to leverage USD 5-8 million in private investment per year. Such figures would suggest the initiative has been highly cost effective in inducing pollution abatement in agricultural operations in Pennsylvania. However, as tax credits are issued based on the introduction of standard, pre-approved practices and products (an outcome-based approach), the impact on pollution will vary from site to site and over time, reducing static cost-efficiency. Additionally, it is not clear to what extent such investments were previously prevented by misaligned incentives, or by issues such as information failures, with further consequences for cost-efficiency. The eligibility of only pre-approved BMPs also reduces the incentive to innovate, preventing dynamic cost-efficiency.

Instrument feasibility. By allowing any tax paying entity to fund (or 'sponsor') improvements, potential issues of access to capital are tackled. An annual cap on the number and value of credits available is set each year, allowing the ability to learn from previous years.

6.2.6 The Agriculture and Environment Programme for Vittel area: an example of payment for ecosystem services (PES)

The bottled water brand 'Vittel' sources and bottles its water from an aquifer in the village of the same name, in northeast France. To maintain its status as 'natural mineral water' the nitrate concentration must remain stable, not exceed 15mg/l, not contain pesticides, and not be treated at any stage in the process. In the early 1980s, the intensification of agriculture in the local catchment threatened to violate these conditions, and by extension, the continuation of the Vittel operation. Various options were considered to mitigate this threat, including purchasing the agricultural land causing the problem and using legal action to force the agricultural operations to alter their practices. However, legal and practical issues rendered such options infeasible. As such, in 1989, a system to incentivise farmers to voluntarily alter their practices was decided upon. Extensive research was undertaken to determine the relationship between farming practices and the nitrate rate in the aquifer, the practices available to minimise excessive nitrogen release, and the incentives necessary to encourage farmers to adopt these practices (beyond legal requirements) as a proxy for the provision of water with low nitrate concentrations (an outcome-based approach). Based on this, an incentive package was created. Farmers must agree to eliminate corn crops, ban pesticides, limit the use of artificial fertiliser, compost all animal waste, limit livestock intensity and ensure high standards for buildings (Depres et al., 2005). In exchange, Vittel would provide up to EUR 150 000 per farm to cover the cost of new equipment and building modernisation, cover the cost of labour for the application of compost in fields, provide free technical advice and assistance, and an average support of EUR 200/ha/year for five years to ensure income during practice transition (Perrot-Maitre, 2006).

Instrument effectiveness. The Agriculture and Environment Programme for Vittel area (AGREV) has been widely recognised as a success. By 2004, all farms in the area had joined the scheme, protecting 92% of the at-risk area, with the nitrogen levels in the aquifer remaining stable. The causal links between existing agricultural practices and nitrogen pollution in the local catchment were well researched and understood, in order to determine appropriate measures to promote. However, it cannot be known to what extent the instrument has reduced nitrogen concentrations in the aquifer below those that would have resulted in the absence of the scheme. Payments are not conditional upon reduced nitrogen pollution (impact), but on the purchase and implementation of given technologies and practices (outcome), reducing both static and dynamic cost-efficiency (as the impact on nitrogen release will likely differ according to variables including location and time).

Instrument cost-efficiency. It is estimated that Vittel spent EUR 24.25 million over the first seven years of the AGREV programme - a rate of EUR 980/ha, and EUR 1.52 per m^3 of bottled water produced. The cost effectiveness of the scheme in the long run is shown by the continued profitability of the Vittel brand (Perrot-Maitre, 2006).

Instrument feasibility. A critical component of success of the scheme was the effort made to understand the choices faced and made by each operation within the catchment individually (including non-cost issues such as inheritance laws), and long-term, open dialogue to forge trust and mutual understanding. The final package received by farmers was negotiated on an individual basis to address their particular needs and concerns. Although this increased feasibility, it likely increased transaction costs significantly (Perrot-Maitre, 2006).

6.2.7 Australia's 'FERTCARE': an example of information measure

Agriculture is central to many environmental debates in Australia (Drew, 2007), with the extensive use of manufactured fertiliser in Australia of significant concern, particularly surrounding consequential eutrophication in inland and coastal waterways (of which a decline in water quality of the latter threatens the health of the Great Barrier Reef to the northeast of the country). In 2004 the FERTCARE programme (the 'Programme', hereafter), a joint initiative between the Australian Fertiliser Services Association (AFSA), Fertiliser Australia, and the Departments of Environment and Heritage (DEH) and Agriculture, Fisheries and Forestry (DAFF), was launched as the centrepiece instrument for tackling these issues. Farmer advisory services, sellers of fertiliser spreading machines and fertiliser manufacturers seeking to minimise the environmental damage from manufactured fertiliser application can undertake FERTCARE training. Advisors can take a step beyond FERTCARE training and be recognised as FERTCARE Accredited Advisors (FAA). FAA advice on soil and fertiliser management should be based on high quality soil and/or plant testing methodology and on laboratories applying good practice and accepted science in Australia. FAA advisors are subject to a biennial audit. Sellers of fertiliser spreaders can apply for FERTCARE "Accu-Spread" certification. To display the FERTCARE "organisation" logo, fertiliser manufacturers must comply with the objectives of the fertiliser industry in terms of staff training. FERTCARE has been constantly updated and won a Business and Higher Education Round Table (B-HERT) award in 2012.

Instrument effectiveness. The Programme aims to train all eligible people. As of 2016, twelve years after the start of the programme, only 76% of the eligible staff in respondent companies had successfully completed FERTCARE training (Fertiliser Australia, 2016). Whilst this might suggest the Programme has been ineffective in achieving its objectives, it is likely that staff turnover and failure for some persons to successfully complete the training mean that 100% training is infeasible (ibid). Eligible staff turnover continues to be an issue facing fertiliser businesses. Additionally, in 2016, only 74 spreading machines were Accu-Spread certified, which is probably low in percentage although there is no reliable data on the number of spreading machines in Australia. In 2016, Australia had 256 FAA advisors and 11 FERTCARE "organisation" businesses. The 'impact' (reduced nitrogen pollution) as a result of the Programme has not been evaluated and is not directly monitored; meaning the environmental effectiveness of the instrument cannot be directly assessed. However, the 'outcome' (induced practice changes on farms and other sites) may be used as a proxy to some extent. Cummins (2016) find in its evaluation of the FERTCARE Carbon Farming Extension Project (FCFEP), which ran from 2013 to 2016, that FAA advisors were giving better advice to farmers based on their improved knowledge of emissions reductions, carbon storage and government policy. Overall, 59% of survey participants indicated that FERTCARE nitrogen use efficiency training materials were very relevant to them as advisors, and that figure rises to 69% for those who have completed the training compared to 43% of those who have not. Therefore, whilst the Programme appears to be relatively successful in achieving its direct objectives of widespread training and the provision of high-quality advice to farmers, the overall environmental impact is unclear.

Instrument cost-efficiency. In 2005, fertiliser sales in Australia totalled nearly AUD 2.5 billion (USD 1.9 billion). Drew (2007) estimated the cost to the fertiliser industry of achieving the targets discussed above is likely to be around AUD 4 million (USD 3 million). As such, the Programme represents a relatively low, but still significant cost to the industry.

Instrument feasibility. The potential threat of future regulatory action may have been a factor in achieving acceptance among the fertiliser industry. However Drew (2007) identifies other factors contributing to its feasibility, including:

- Clear statements of the issues of fertiliser overuse by reputable parties
- Identifications of positive implications for the industry from the Programme
- Funding support from the Australian Government for initial establishment
- The commitment to 100% compliance encouraged all participants to invest in the scheme
- The 'all encompassing' approach to training to suit all levels of role complexity
- The involvement of other stakeholders, particularly from the public sector
- The use of an external qualifications framework to manage accreditation and record keeping.

6.2.8 Chesapeake 2000 Programme: an example of voluntary scheme

The Chesapeake Bay is a significant source of economic activity (including fishing, tourism and shipping) for the Mid-Atlantic United States, the development of which over the past several decades has caused degradation (eutrophication) of the Bay's waters, impairing further economic growth. In response, in 2000 the states of Maryland, Virginia, Pennsylvania, the District of Colombia and the federal government, building on an existing partnership, created the Chesapeake 2000 Programme – a voluntary agreement containing over 100 overarching goals, including removal of the Bay from the US Environmental Protection Agency (USEPA)'s 'impaired waters' list, and more specific actions, such as protecting and restoring 114 000 acres (46 000 hectares) of submerged aquatic vegetation (vegetation that grows to the surface of, but does not emerge from, shallow water), support the restoration of key rivers and tributaries, and developing and promoting wastewater treatment options to be (mostly) achieved by 2010 (Cramer, 2014; USEPA and USDA, 2006).

Instrument effectiveness. Cramer (2014) finds that whilst nitrogen loads in the Bay were around 13% lower between 2000 and 2010 than between 1990 and 2000, statistical analysis determines that the trend is not significant (a result that also stands for phosphorus loads). However, when controlled for hurricane and tropical storm events, the programme does appear to have been significant in reducing nutrient loads, with a total reduction of 40 million tonnes of nitrogen between 1990 and 2010 (and 1 million tonnes of phosphorus). However, Chesapeake Bay remained on the list of 'impaired waters', and by the end of 2010, became subject to TMDL regulations by the USEPA (see case study on the Chesapeake Bay Watershed in Chapter 3.).

Instrument cost-efficiency. Ex ante estimates suggested a total projected cost of USD 18.7 billion to achieve all the goals of Chesapeake 2000, compared to an 1989 estimate of the value of the Bay at USD 678 billion (in 1987 dollars) (CBC, 2003). The value of committed resources by the termination of the Programme in 2010 is not clear, although even if it were, given the multitude of overlapping objectives, the specific cost-efficiency of achieved nitrogen pollution abatement through the Programme would be difficult to disaggregate and determine.

Instrument feasibility. Although the Programme was backed up by a credible threat of regulation (evidenced by the eventual imposition of TMDL regulations), other factors are likely to have combined to prevent more significant reduction in nutrient loads. For example, whilst some goals were highly specific in their objective, around half were unquantifiable, producing ambiguity. Additionally, provisions to prevent free-riding by each of the signatories were also lacking, along with positive incentives (e.g. visible participant benefits) to achieve the objectives of the agreement (Cramer, 2014).

Notes

[1] It must be noted that as individual instruments (under a given definition) and their combinations may hold significantly varied specific characteristics and operate under very different contextual conditions, the assessment performed by Drummond et al., 2016, 2015a and 2015 b can be viewed as indicative only.

[2] The 'polluter pays' principle is defined as 'the principle according to which the polluter should bear the cost of measures to reduce pollution according to the extent of either the damage done to society or the exceeding of an acceptable level (standard) of pollution' (OECD Glossary of Statistical Terms).

[3] The pollution haven hypothesis, or pollution haven effect, is the idea that polluting industries will relocate to jurisdictions with less stringent environmental regulations.

[4] In opposition to the 'polluter pays' principle, the 'beneficiary pays' principle states that those who benefit from an action should contribute to the cost of that action (Hatfield-Dodds, 2006).

[5] 'Feebate' instruments (or 'bonus-malus') levy a tax or charge ('fee') on activities or products below a given environmental threshold performance or limit, and provide financial support to activities or products that exceed this performance level ('rebate'). A key example of a revenue-neutral 'feebate' instrument is the Swedish refund emission payment for NO_x (OECD, 2013).

[6] Regulatory capture occurs when a regulatory agency, created to act in the public interest, furthers the interests of groups that dominate the industry or sector it is charged with regulating.

[1] Except for administration costs, amounting to around 0.7% of total revenue (OECD, 2013).

[2] Due to the broad revenue neutrality of the instrument, relative product prices did not change, and therefore demand for such products did not alter. Moreover, as product prices do not change, the occurrence of negative distributional impacts is also prevented (OECD, 2013).

[3] A hotspot may be actively created, for example through the concentration of tradable permits in space, time or sectors, or simply be a location, time or sector in which the marginal cost of pollution is particularly elevated, and not actively addressed by an instrument or instrument combination.

[4] Most often applied in the context of increasing energy efficiency, the Rebound Effect occurs when increasing production efficiency reduces the cost per unit of service delivered, increasing consumption in accordance with the price elasticity of demand (Berkhout et al., 2000).

[5] The externality is targeted directly. Also known as a 'performance-based' approach.

[6] The externality is targeted indirectly, by the prohibition or mandating of certain practices. Also known as a 'technology-based' approach.

[7] Broadly speaking, both impact-based and outcome-based approaches may be applied to both upstream and downstream actors.

[8] The evidence suggests that for a well-known feebate example, the French Bonus-Malus system on new cars that aimed to reduce CO_2 emissions, a pivot rate positioned too high, coupled with overly generous rebated, CO_2 emissions actually increased.

[9] Government (or other public body) is the 'buyer' of public good ecosystem services (e.g. CO_2 sequestration).

[10] Private actors are the 'buyers'; with ecosystem services 'purchased' delivering (largely) private benefit.

[11] The targeted regulation complements existing Danish nitrogen regulations, including those contained within the Food and Agriculture Package, adopted in December 2015.

[12] As opposed to a forward auction, a reverse auction requires credit sellers compete to obtain business from the buyer. In this case, the MCD accepts bids from SWCDs, and then sets the maximum bid value after the first round of submissions.

[13] WWTPs that engage in the system prior to the introduction of binding legislation require a single credit per pound of nutrient in excess of future binding limits, whilst those that purchase credits after the introduction of legislation will require up to three credits.

[14] Including Personal Income Tax, Corporate Net Income Tax, Capital Stock and Franchise Tax, Bank Shares Tax, Title Insurance Company Premiums Tax, Insurance Premiums Tax and Mutual Thrift Institutions Tax (PSCC, 2011).

References

Arimura, T.H. and K. Iwata (2006), "Inefficiency in Command-and –Control Approach Toward Adoption of Low Emission Vehicles: Japanese Experience of Air Pollution from Diesel Trucks", Paper presented at 3rd World Congress of Environmental and Resource Economists, Kyoto, Japan.

Baumol, W.J. and W.E. Oates (1988), *The Theory of Environmental Policy*, 2nd Ed., Cambridge University Press.

Bennear, L.S. and R.N. Stavins (2007) "Second-Best Theory and the Use of Multiple Policy Instruments", *Environmental Resource Economics*, 37.

Berkhout, P.H.G. et al. (2000), "Defining the Rebound Effect", *Energy Policy*, 28(6-7).

Bonilla, J. et al. (2014), "Diffusion of NO_x Abatement Technologies in Sweden", Working Papers in Economics N°585, Department of Economics, University of Gothenburg.

Chesapeake Bay Commission (CBC) (2003), *The Cost of a Clean Bay: Assessing Funding Needs Throughout the Watershed*, Chesapeake Bay Commission, Annapolis, United States.

Colls, J. and A. Tiwary (2010), *Air Pollution: Measurement, Modeling and Mitigation*, 3rd Ed., Oxford, Routledge.

Cramer, S. (2014), "An Examination of Levels of Phosphorus and Nitrogen in the Chesapeake Bay Before and After the Implementation of the Chesapeake 2000 Program", *The Public Purpose*, 12.

Cummins, T. (2016), "FERTCARE Carbon Farming Extension Project Second Survey – Advisor Knowledge and Attitudes", Tim Cummins and Associates.

Danish Economic Councils (2018), "Economy and Environment, 2018", Summary and Recommendations, De Økonomiske Råd.

D'Haultfoeuille, X. et al. (2014), "The Environmental Effect of Green Taxation: The Case of the French *Bonus/Malus*", *The Economic Journal*, 124(578).

De Haan, P. et al. (2009), "How Much Do Incentives Affect Car Purchase? Agent-Based Microsimulation of Consumer Choice of New Cars – Part II: Forecasting Effects of Feebates Based on Energy-Efficiency", *Energy Policy*, 37(3).

Depres, C. et al. (2005), *On Coasean Bargaining with Transaction Costs: The Case of Vittel*, CESAER Working Paper 2005/3.

Drew, N. (2007), "Fertcare – Putting Best Practice into Stewardship", in: *Fertiliser Best Management Practices: General Principles, Strategy for their Adoption and Voluntary Initiatives vs Regulations*, International Fertiliser Industry Association, Paris.

Drummond, P. et al. (2016), "Policy Instruments and Combinations to Manage the Unwanted Release of Nitrogen into Ecosystems – Effectiveness, Efficiency and Feasibility", paper presented to the Working Party on Biodiversity, Water and Ecosystems at its meeting on 11-12 May 2016, ENV/EPOC/WPBWE(2016)5.

Drummond, P. et al. (2015a), "Policy Instruments to Manage the Unwanted Release of Nitrogen into Ecosystems – Effectiveness, Cost-Efficiency and Feasibility", paper presented to the Working Party on Biodiversity, Water and Ecosystems at its meeting on 19-20 February 2015, ENV/EPOC/WPBWE(2015)8.

Drummond, P. et al. (2015b), "Policy Instrument Combinations to Manage the Unwanted Release of Nitrogen into Ecosystems – Effectiveness, Cost-Efficiency and Feasibility", paper presented to the Working Party on Biodiversity, Water and Ecosystems at its meeting on 21-22 October 2015, ENV/EPOC/WPBWE(2015)14.

Ecotec (2001), "Study on the Economic and Environmental Implications of the Use of Environmental Taxes and Charges in the European Union and its Member States", Ecotec, Brussels.

Fertiliser Australia (2016), "Sustainability and Stewardship 2016", www.fertilizer.org.au/Portals/0/Documents/Reports/Sustainability%20and%20Stewardship%202016.pdf?ver=2018-01-22-151251-473.

Greene, D.L. et al. (2005), "Feebates, Rebates and Gas-Guzzler Taxes: A Study of Incentives for Increased Fuel Economy", *Energy Policy*, 33(6).

Hasunuma, H. et al. (2014), "Decline of Ambient Air Pollution Levels due to Measures to Control Automobile Emissions and Effects on the Prevalence of Respiratory and Allergic Disorders among Children in Japan", *Environmental Research*, 131.

Hatfield-Dodds S. (2006), "The Catchment Care Principle: A New Equity Principle for Environmental Policy, with Advantages for Efficiency and Adaptive Governance", *Ecological Economics*, 56(3).

Höglund-Isaksson, L. and T. Sterner (2009), *Innovation Effects of the Swedish NO_x Charge*, OECD Global Forum on Eco-Innovation, 4-5 November 2009, OECD, Paris.

Iwata, K. and T.H. Arimura (2008), "Economic Analysis of a Japanese Air Pollution Regulation: An Optimal Retirement Problem under Vehicle Type Regulation in the NO_x-Particulate Matter Law", Discussion Paper 08-15, Resources for the Future, Washington D.C.

Iwata, K. et al. (2014), *The Effectiveness of Vehicle Emission Control Policies: Evidence from Japanese Experience*, Working Paper E-77, Tokyo Centre for Economic Research, Tokyo.

Johnson, K.C. (2006), "Feebates: An Effective Regulatory Instrument for Cost-Constrained Environmental Policy", *Energy Policy*, 34(18).

Kieser and Associates (2004), "Preliminary Economic Analysis of Water Quality Trading Opportunities in the Great Miami River Watershed, Ohio", Kieser and Associates, Kalamazoo, United States.

Kolstad, C.D. (2011), *Intermediate Environmental Economics*, 2nd ed., Oxford, Oxford University Press.

Miami Conservancy District (MCD) (2014), *Water Quality Credit Trading Program: A Common Sense Approach to Reducing Nutrients*, newserver.miamiconservancy.org/water/documents/WQCTPfactsheet2014FINAL_000.pdf.

Naturvårdsverket (2014), *Ändring av kväveoxidavgiften för ökad styreffekt: Redovisning av ett regeringsuppdrag*, Rapport 7747, Naturvårdsverket, Stockholm.

Newburn, D.A. and R.T. Woodward (2012), "An Ex Post Evaluation of Ohio's Great Miami Water Quality Trading Program", *Journal of the American Water Resources Association*, 48(1).

OECD (2013), "The Swedish Tax on Nitrogen Oxide Emissions: Lessons in Environmental Policy Reform", *OECD Environment Policy Papers*, No. 2, OECD Publishing, Paris, doi.org/10.1787/5k3tpspfqgzt-en.

OECD (2006), *The Political Economy of Environmentally Related Taxes*, OECD Publishing, Paris, doi.org/10.1787/9789264025530-en.

OECD (2002), *OECD Environmental Performance Reviews: Japan 2002*, OECD Environmental Performance Reviews, OECD Publishing, Paris, doi.org/10.1787/9789264175334-en.

Pattanayak, S.K. et al. (2010), "Show Me the Money: Do Payments Supply Environmental Services in Developing Countries?", *Review of Environmental Economics and Policy*, 4(2).

Perrot-Maitre, D. (2006), *The Vittel Payments for Ecosystem Services: A 'Perfect' PES Case?*, International Institute for Environment and Development, London.

Pennsylvania State Conservation Commission (PSCC) (2011), *PA REAP Resource Enhancement and Protection FY 2009-10 and FY 2010-11 Annual Report*, Pennsylvania Department of Agriculture, United States.

Shortle, J. (2012), "Water Quality Trading in Agriculture", paper prepared for the OECD Joint Working Party on Agriculture and the Environment, COM/TAD/CA/ENV/EPOC(2010)19/FINAL, OECD, Paris.

SOU (2017), "Brännheta skatter! Bör avfallsförbränning och utsläpp av kväveoxider från energiproduktion beskattas?", Betänkande av Förbränningsskatteutredningen, Statens Offentliga Utredningar, SOU 2017:83, Stockholm.

Sterner, T. and B. Turnheim (2008), "Innovation and Diffusion of Environmental Technology: Industrial NO_x abatement in Sweden under Refunded Emission Payments", Discussion Paper, Resources for the Future, Washington D.C.

Tomich, T.P. et al (2004), "Environmental Services and Land Use Change in Southeast Asia: From Recognition to Regulation or Reward?", *Agriculture Ecosystems and Environment*, 104.

USEPA and USDA (2006), "Evaluation Report: Saving the Chesapeake Bay Watershed Requires Better Coordination of Environmental and Agricultural Resources", US Environmental Protection Agency and US Department of Agriculture, Washington D.C

Annex A. Basic facts on nitrogen

A.1 The nitrogen cycle

Conversion of the highly stable ("inert") dinitrogen (N_2) molecule to biologically available ("reactive") nitrogen, a process called "fixation", is difficult. Fixation is achieved in soil and water by specialised bacteria which can reduce atmospheric dinitrogen to ammonia (NH_3) or ammonium (NH_4^+) (Figure A.1).[1] Microorganisms early on in the Earth's history developed the ability to use enzymes to produce or "fix" NH_4^+ from dinitrogen, possibly because the availability of nitrogen through abiotic routes were biologically limiting (McRose et al., 2017). Nitrogen-fixing prokaryotes (bacteria and archaea) live in water (e.g. cyanobacteria), in soil (e.g. Azotobacter), in association with plants (e.g. Azospirillum), or in symbiosis with leguminous plants such as peas, clover and soyabeans (e.g. Rhizobium). In the latter case, the prokaryote shares the nitrogen with the plant; in exchange, the plant supplies the prokaryote with a source of carbon and energy for growth.

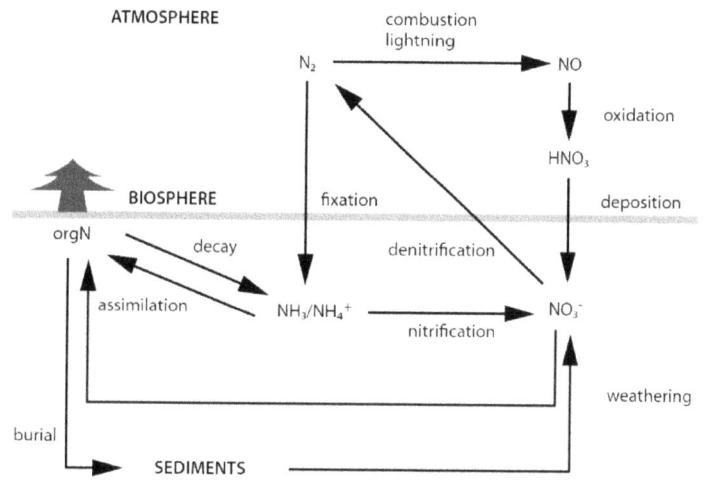

Figure A.1. The nitrogen cycle: major processes

Note: Ammonia (NH_3) is highly soluble in water. When dissolved in water, a portion of NH_3 reacts to form ammonium (NH_4^+) as a function of the acidity of the water, i.e. its pH ("potential of hydrogen"). pH refers to the concentration of hydrogen ions (H^+) in water. A decrease in pH results in an increase in NH_4^+ and a decrease in NH_3.
Source: Jacob (1999).

The NH_3/NH_4^+ is "assimilated" as organic nitrogen by the bacteria or by their host plants, which may in turn be consumed by animals. Eventually these organisms excrete the nitrogen or die; the organic nitrogen is eaten by bacteria and mineralised to NH_3/NH_4^+,[2] which may then be assimilated by other organisms.

Bacteria may also use NH_3/NH_4^+ as a source of energy by oxidising it to nitrite (NO_2^-) and on to nitrate (NO_3^-). This process ("nitrification") requires oxygen (aerobic conditions).[3] NO_3^- is highly mobile in soil and is readily assimilated by plant and bacteria, providing another route for formation of organic nitrogen.

When oxygen is depleted in water or soil (anaerobic conditions), bacteria may use NO_3^- as an alternate oxidant to convert organic carbon to carbon dioxide (CO_2). This process ("denitrification") converts NO_3^- to dinitrogen (N_2) and thus returns nitrogen from the biosphere to the atmosphere.[4] Denitrification may generate nitrous oxide (N_2O).

An additional pathway for fixing dinitrogen is by high-temperature oxidation of dinitrogen to nitric oxide (NO) in the atmosphere during combustion (e.g. forest fire) or lightning, followed by atmospheric oxidation of NO to nitric acid (HNO_3) which is water-soluble and scavenged by rain.

Nitrogen is transferred to the lithosphere by burial of dead organisms (including their nitrogen) in the bottom of the ocean. These dead organisms are then incorporated into sedimentary rock. Eventually the sedimentary rock is brought up to the surface of the continents and eroded, liberating the nitrogen and allowing its return to the biosphere. Rock (geologic) nitrogen comprises a potentially large pool of nitrogen (Holloway and Dahlgren, 2002). Rock nitrogen concentrations range from trace levels (<200 mg N per kg) in granites to more than 1 000 mg N per kg in sedimentary rocks. Nitrate deposits accumulated in arid and semi-arid regions are also a large potential pool.

A.2   The nitrogen problem in brief

According to anthropogenic and natural nitrogen flux estimates compiled by Battye et al., 2017, human nitrogen production is estimated to have increased fivefold in the last half-century (since 1960). According to Bleeker et al., 2013, human emissions of nitrogen oxides (NO_x)[5] and NH_3 to the atmosphere have increased about fivefold since pre-industrial times. Atmospheric nitrogen deposition in large regions of the world exceeds natural rates by an order of magnitude. Much is deposited in nitrogen-limited ecosystems, leading to unintentional fertilisation and loss of biodiversity.

Still according to Bleeker et al., 2013, the transfer of nitrogen from terrestrial to coastal systems has doubled since pre-industrial times. As with terrestrial ecosystems, many of the coastal ecosystems are nitrogen-limited, such that abundance in nitrogen leads to algal blooms and a decline in the quality of aquatic ecosystems.

In addition, nitrogen has direct and indirect effects on climate change, being itself a greenhouse gas (GHG) and influencing emissions and the uptake of other GHGs such as CO_2. It also enhances ground-level ozone (GLO) formation, depletes stratospheric ozone, increases soil acidification and stimulates the formation of particles in the atmosphere, all of which have negative effects on people and the environment.

The social cost of nitrogen impacts appears to be largely underestimated.[6] Considering only the health impact of air pollution by nitrogen, the social cost is already in the hundreds of billions of USD. This is because nitrogen represents a significant part of urban pollution with fine particles (PM$_{2.5}$)[7] whose health cost (premature deaths) is estimated at some USD 1.8 trillion in OECD countries and USD 3.0 trillion in BRIICS countries (Roy and Braathen, 2017). When adding GLO pollution, of which nitrogen is also a precursor, health costs amount to some USD 1.9 trillion in OECD countries and USD 3.2 trillion in BRIICS countries (ibid).

To this must be added the health cost of water pollution by nitrogen as well as the cost of nitrogen pollution on ecosystems and climate. Keeler et al., 2016 made estimates for the state of Minnesota in the United States (see Chapter 2.).

Global warming could make nitrogen pollution worse (Conniff, 2017). Nitrogen availability changes in response to climate change, generally increasing with warmer temperatures and increased precipitation. For example, Sinha et al., 2017 projects that climate change–induced precipitation changes alone will increase runoff nitrogen in U.S. waterways by 19% on average over the remainder of the century under a business as usual climate scenario. Toxic blue-green algae (or cyanobacteria) blooms, fuelled by nitrogen pollution, are being exacerbated by warmer temperatures and increased rainfall associated with climate change (Paerl et al., 2016).

However, there is not yet a thorough understanding of the complex interactions between climate change and the nitrogen cycle (e.g. in the case of seasonally inundated environments) and what the net effect might be under different scenarios of future anthropogenic activity. In particular, more research is needed on the joint impact of climate change and nitrogen on plant diversity. For example, an experiment in arid habitat in southern California has shown that nitrogen combined with changing precipitation can result in a community of native shrubs to shift to non-native grasses (Rao and Allen, 2010).

A.3 Supplementary information on nitrogen impacts

Chapter 1 provides an overview of the main externalities associated with excess nitrogen in the environment. The following sections provide additional factual information. Section A.3.1 focuses on the troposphere.[8]

A.3.1 Air quality

Nitrogen dioxide (NO$_2$) and, to a lesser extent, ammonia (NH$_3$), are directly harmful to human health. Nitric oxide (NO) is not considered to be hazardous to health at typical ambient concentrations, but at high concentrations NO$_2$ is toxic, which is why it is classified by the World Health Organisation (WHO) as a hazardous atmospheric pollutant.

Parts of the secondary particulate matter (PM) are formed in the atmosphere from nitrogen oxides (NO$_x$) and NH$_3$ precursors. NO$_x$ is also an essential precursor to the formation of ground-level ozone (GLO) (in the presence of sunlight).

Nitrogen dioxide (NO$_2$)

There is strong evidence of respiratory effects (asthma exacerbation) following short-term NO$_2$ exposures, typically minutes to hours (in part, following USEPA, 2016a and 2017). In addition to the effects of short-term exposures, there is likely to be a causal relationship between long-term NO_2 exposures and respiratory effects, based on the evidence for asthma development in children. People with asthma, as well as children and the elderly are generally at greater risk for the health effects of NO_2.

Ammonia (NH₃)

Short-term inhalation exposure to high levels of NH_3 in humans can cause irritation and serious burns in the mouth, lungs, and eyes (in part, following USEPA, 2016b). Chronic exposure to airborne NH_3 can increase the risk of respiratory irritation, cough, wheezing, tightness in the chest, and impaired lung function. In animals, breathing NH_3 at sufficiently high concentrations can similarly result in effects on the respiratory system. Animal studies also suggest that exposure to high levels of ammonia in air may adversely affect other organs, such as the liver, kidney, and spleen.

Particulate matter (PM)

PM pollution stands for a mixture of solid particles and liquid droplets found in the air (in part, following USEPA, 2013a). These particles come in many sizes and shapes and can be made up of hundreds of different chemicals. Some (primary PM) are emitted directly from a source, such as construction sites, unpaved roads, fields, chimneys or fires. Most form in the atmosphere as a result of chemical reactions; these are secondary PMs such as ammonium nitrate (NH_4NO_3).

PMs can be inhaled and cause serious health problems. They can get deep into lungs, and some may even get into bloodstream. Exposure to PM increases the risk of premature death in people with heart or lung disease, nonfatal heart attacks, irregular heartbeat, aggravated asthma, decreased lung function, and increased respiratory symptoms, such as irritation of the airways, coughing or difficulty breathing. People with heart or lung diseases, children, and older adults are the most likely to be affected by particle pollution exposure. The health effects are associated with long- and short-term exposures. The risk of mortality is higher for long-term exposure. The risk of morbidity for short-term exposure is higher for cardiovascular than respiratory diseases.

Ground-level ozone (GLO)

Breathing GLO (or "bad" ozone) can trigger reduced lung function, increased respiratory symptoms and pulmonary inflammation, particularly for children, the elderly, and people with asthma or other lung diseases, and other at-risk populations (in part, following USEPA, 2015). Health effects are associated with long- and short-term exposures. Research also indicates that GLO exposure can increase the risk of premature death from heart disease, although more research is needed to understand how GLO may affect the heart and cardiovascular system. GLO is most likely to reach unhealthy levels on hot sunny days in urban environments, but can still reach high levels during colder months.

GLO can also have harmful effects on sensitive vegetation and ecosystems, including crops, in particular during the growing season. When sufficient GLO enters the leaves of a sensitive plant, it can reduce photosynthesis, slow the plant's growth and increase the plants' risk of disease. The effects of GLO on individual plants can then have negative impacts on ecosystems, including loss of species diversity, changes to species composition, changes to habitat quality and changes to water and nutrient cycles.

A.3.2 Greenhouse balance

Three forms of nitrogen have a direct impact on the greenhouse effect (Erisman et al., 2011). First is nitrous oxide (N_2O), which has a strong global warming potential over a 100-year time scale (GWP_{100} close to 265).[9] GWP_{100} conversion factors have changed in successive reports of the Intergovernmental Panel on Climate Change. There are various reasons for this, including temperature of the atmosphere and the increasing concentration of the greenhouse gases (GHG) themselves. These conversion factors do not include the indirect effects that N_2O has on carbon-cycle feedbacks, such as the impact the warming N_2O causes will have on how much carbon the world's oceans and forests will be able to store in the future. These effects are thought to be large – their inclusion would increase the 2013 conversion factor for N_2O from 265 to 298 (IPCC, 2013).

Second is nitrogen oxides (NO_x) emissions which contribute to (i) formation of ground-level ozone (GLO); (ii) a decrease of methane (CH_4); and, (iii) formation of nitrate aerosols. CH_4 and GLO are, respectively, the 2nd and 3rd most important GHGs after carbon dioxide (CO_2).[10] Nitrogen-containing aerosols have a cooling effect, both direct and indirect (through cloud formation).[11] CH_4 has both a warming effect (as a GHG) and a cooling effect (through formation of aerosol in the troposphere). The net effect of all three NO_x-related contributions is cooling. Third is ammonia (NH_3) emissions which contribute to aerosol formation and a cooling effect.

Nitrogen also impacts indirectly on climate change (Erisman et al., 2011). First, microbial ammonification of dissolved organic nitrogen (DON) in soils (litter decomposition) contributes to soil respiration and hence CO_2 emissions. Second, nitrogen increases plant productivity and hence CO_2 uptake in terrestrial ecosystems, except in situations where it accelerates organic matter breakdown, thereby increasing release of CO_2. Third, nitrogen increases marine productivity and hence CO_2 uptake in oceans, except in situations where it causes ocean acidification. Fourth, GLO reduces plant productivity, and hence CO_2 uptake by plants.

If considered in all its forms, and not only N_2O, and in all its direct and indirect effects, Erisman et al, 2011 estimated that there was no significant net effect of nitrogen on overall radiative balance. However, this estimate does not take into account the fact that the effects of NO_x and NH_3 have a relatively short life span whereas N_2O persists in the atmosphere for about a century. Measures that reduce short-lived aerosols but do not address long-lived N_2O would, for example, increase the net warming effects of nitrogen.

A.3.3 Water quality

Anthropogenic increase of nitrogen in water poses direct threats to human and aquatic ecosystems (fresh and marine waters). High nitrate/nitrite (NO_3^-/NO_2^-) levels in drinking water may cause a potentially fatal blood disorder in infants under six months of age called methemoglobinemia or "blue-baby" syndrome. With this disorder there is a reduction in the oxygen carrying capacity of blood, which can cause shortness of breath and a blueness of the skin of infants or even lead to the infant's death.

The US Environmental Protection Agency (USEPA) has not mandated a Maximum Contaminant Level (MCL) for ammonia (NH_3). However, it has been known, since early in this century, that NH_3 is toxic to fish and that the toxicity increases with increasing pH and temperature of the water (USEPA, 2013b). When NH_3 is present in water at high enough levels, it is difficult for aquatic organisms to sufficiently excrete the toxicant, leading to toxic build-up in internal tissues and blood, and potentially death. In 2013, USEPA has issued a Final Ammonia Criteria for the toxic effect of NH_3 for aquatic life.

In aquatic ecosystems, nutrient enrichment (eutrophication) is responsible for algal blooms (including toxic algal blooms[12]) on the surface. This can lead to the reduction or even the disappearance of oxygen and thus of fish in the deep waters that become "dead zones".

Research continues to reveal nitrogen-related impacts on health and aquatic ecosystems. For example, Zhang et al., 2015 found a link between the proliferation of toxic blue-green algae and the risk of death from liver disease. Another example relates to coral. Corals are adapted to thrive in the sun-lit, nutrient-poor waters of tropical oceans[13] thanks to their intimate relationship with microscopic algae and nitrogen-fixing microbes (called diazotrophs). Photosynthesis of algae provides the coral animal with a source of carbon and energy, while diazotrophs provide nitrogen for metabolism and growth. In this relationship, corals regulate the algal growth by limiting their access to nitrogen (Pogoreutz et al., 2017). However, sugar-enriched discharges in coastal waters (e.g. from wastewater) feed the diazotrophic activity, which means they fix more nitrogen. This excess nitrogen available for algae causes the breakdown of coral-algae symbiosis and triggers bleaching (ibid).[14]

A.3.4 Ecosystems and biodiversity

Species and communities most sensitive to chronically elevated nitrogen deposition are those that are adapted to low nutrient levels, or are poorly buffered against acidification.[15] In the United Kingdom, nitrogen affects many threatened vascular plant, bryophyte and lichen species as well as certain fungal groups (Plantlife and Plant Link UK, 2017). Early evidence also suggests that habitat changes resulting from nitrogen deposition may also affect other taxonomic groups such as insects and birds, although further research is required (ibid). In addition, after an initial increase, the productivity of forests and grasslands is reduced beyond a certain threshold of nitrogen loading.

Hernández et al., 2016 showed that 78 of the 1 400 species of invertebrates, vertebrates and plants listed under the US Endangered Species Act (approximately 6%) were affected by the direct toxicity of nitrogen; eutrophication of their habitat or the spread of non-native plant species. Simkin et al., 2016 found that nitrogen deposition exceeded critical loads for loss of plant species richness in 24% of the 15 136 sites they examined in the United States. Those sites included woodland and grassland sites across the United States. They found that the effects of nitrogen deposition on plant species richness was more pronounced in acidic versus neutral or basic soil and in dry versus wet climates.

SRU, 2015 reveals that in 2009, 48% of Germany's natural and semi-natural terrestrial ecosystems were affected by eutrophication and 8% by acidification.

Jones et al., 2014 estimated the impact of declines in nitrogen deposition on the value of six ecosystem services in the United Kingdom: two provisioning (timber and grassland production), two regulating (CO_2 sequestration and reduction of N_2O emissions) and two cultural services (recreational fishing and appreciation of biodiversity). They found a net benefit with reduced emissions of N_2O as a GHG and enhanced cultural services outweighing costs (loss of value) of reduced CO_2 sequestration and provisioning services.

Notes

[1] Converting dinitrogen into NH_3/NH_4^+ is the role of "nitrogen-fixing bacteria".

[2] Decomposition of organic matter is the role of decomposers ("ammonifying bacteria").

[3] Converting NH_3/NH_4^+ into NO_2^- and NO_3^- is the role of "nitrifying bacteria".

[4] Returning NO_3^- back into dinitrogen, thereby closing the nitrogen cycle, is the role of "denitrifying bacteria". Denitrifying bacteria comprise more than 150 known species and probably hundreds of unknown species.

[5] Nitrogen oxides (NO_x) include nitric oxide (NO) and nitrogen dioxide (NO_2).

[6] It was estimated within a range of EUR 75–485 billion per year in the EU-27 (Van Grinsven et al., 2013) and USD 81–441 billion per year in the United States (Sobota et al., 2015).

[7] For example, Huang et al., 2017 estimated that nitrate and ammonium aerosols accounted for up to a third of $PM_{2.5}$ emissions measured in Beijing, Tianjin and Shijiazhuang from June 2014 to April 2015. The $PM_{2.5}$ emissions consisted mainly of organic matter (16.0 %–25.0 %), sulphate aerosol (14.4 %–20.5 %), nitrate aerosol (15.1 %–19.5 %), ammonium aerosol (11.6 %–13.1 %) and mineral dust (14.7 %–20.8 %). The nitrogenous aerosol formation pathways are detailed in Chapter 1.

[8] The troposphere starts at Earth's surface and goes up to a height of 7 to 20 km above sea level, depending on latitude and season (and whether it is day or night). Atmospheric layers (i.e. troposphere, stratosphere) have been defined to reflect significant variations in temperature and pressure with altitude.

[9] This means that one tonne of N_2O emitted this year causes 265 times as much heating over the next hundred years as a tonne of CO_2.

[10] Because of its local and short-lived nature, ground-level ozone (GLO) does not have in itself strong global warming effects, but may have a radiative forcing (warming) effect on regional scales; there are regions of the world where GLO has a radiative forcing up to 150% of CO_2. The radiative forcing effect from GLO is strongly height-and latitude-dependent through coupling of GLO change with temperature, water vapour and clouds (Bowman et al., 2013).

[11] Aerosols in the atmosphere scatter and absorb visible radiation, limiting visibility. They affect the Earth's climate both directly (by scattering and absorbing radiation) and indirectly (by serving as nuclei for cloud formation).

[12] Zhang et al., 2015 found evidence of blue-green (toxic) algal blooms in 62% of the 3 100 U.S. counties surveyed.

[13] Much of the world's marine waters lack biologically available nitrogen, which limits photosynthesis in the oceans.

[14] Bleaching occurs when algae that live inside corals and give them their colour are expelled (e.g. due to increased sea temperatures). Bleached corals continue to live but as the algae provide the coral with up to 90% of its energy, after expelling the algae the coral begins to starve and is subject to mortality.

[15] Grassland, heathland, peatland, forest, and arctic/montane ecosystems are recognised as vulnerable habitats in Europe; other habitats may be vulnerable but are still poorly studied (Sutton et al., 2011).

References

Bleeker, A. et al. (2013), "Economy Wide-Nitrogen Balances and Indicators: Concept and Methodology", OECD Working Party on Environmental Information, ENV/EPOC/WPEI(2012)4/REV1.

Bowman, K. W. et al. (2013), "Evaluation of ACCMIP Outgoing Longwave Radiation from Tropospheric Ozone using TES Satellite Observations", *Atmospheric Chemistry and Physics*, 13(8).

Conniff, R. (2017), "The Nitrogen Problem: Why Global Warming Is Making It Worse", *YaleEnvironment360*, 7 August 2017, Yale School of Forestry & Environmental Studies, e360.yale.edu/features/the-nitrogen-problem-why-global-warming-is-making-it-worse.

Erisman, J.W. et al. (2011), « Reactive Nitrogen in the Environment and its Effect on Climate Change », *Current Opinion in Environmental Sustainability*, 3(5).

Hernández, D. L. et al. (2016), "Nitrogen Pollution is Linked to US Listed Species Declines", *BioScience*, 66 (3).

Holloway, J. M. and R. A. Dahlgren (2002), "Nitrogen in Rock: Occurrences and Biogeochemical Implications", *Global Biogeochemical Cycles*, 16(4).

Huang, X. et al. (2017), "Chemical Characterization and Synergetic Source Apportionment of $PM_{2.5}$ at Multiple Sites in the Beijing–Tianjin–Hebei Region, China", *Atmos. Chem. Phys. Discuss.*, 17, doi.org/10.5194/acp-17-12941-2017.

IPCC (2013), *Climate Change 2013: The Physical Science Basis*, Contribution of Working Group I to the Fifth Assessment Report of the Intergovernmental Panel on Climate Change, AR5, Chapter 8, Stocker et al. (eds.), Cambridge University Press, www.ipcc.ch/pdf/assessment-report/ar5/wg1/WG1AR5_Chapter08_FINAL.pdf.

Jacob, D.J. (1999), *Introduction to Atmospheric Chemistry*, Princeton University Press.

Johnson, J.S. et al. (2015), "Evaluating Uncertainty in Convective Cloud Microphysics Using Statistical Emulation", *Journal of Advances in Modeling Earth Systems*, 7(1), doi.org/10.1002/2014MS000383.

Jones, L. et al. (2014), "A Review and Application of the Evidence for Nitrogen Impacts on Ecosystem Services", *Ecosystem Services*, 7, doi.org/10.1016/j.ecoser.2013.09.001.

Keeler, B.L. et al. (2016), "The Social Costs of Nitrogen", *Science Advances*, 2(10).

McRose, D.L. et al. (2017), "Diversity and Activity of Alternative Nitrogenases in Sequenced Genomes and Coastal Environments", *Front Microbiol.*, 8(267), doi.org/10.3389/fmicb.2017.00267.

OECD (2008), *OECD Environmental Outlook to 2030*, OECD Publishing, Paris, doi.org/10.1787/9789264040519-en.

Paerl, H. W. et al. (2016), "Mitigating Cyanobacterial Harmful Algal Blooms in Aquatic Ecosystems Impacted by Climate Change and Anthropogenic Nutrients", *Harmful Algae*, 54, doi.org/10.1016/j.hal.2015.09.009.

Plantlife and Plant Link UK (2017), "We Need to Talk About Nitrogen, The Impact of Atmospheric Nitrogen Deposition on the UK's Wild Flora and Fungi", January 2017.

Pogoreutz, C. et al. (2017), "Sugar Enrichment Provides Evidence for a Role of Nitrogen Fixation in Coral Bleaching", *Glob Change Biol.*, 23(9), doi.org/10.1111/gcb.13695.

Rao, L.E. and E. B. Allen (2010), "Combined Effects of Precipitation and Nitrogen Deposition on Native and Invasive Winter Annual Production in California Deserts", *Oecologia*, 162(4), doi.org/10.1007/s00442-009-1516-5.

Roy, R. and N. Braathen (2017), "The Rising Cost of Ambient Air Pollution thus far in the 21st Century: Results from the BRIICS and the OECD Countries", *OECD Environment Working Papers*, N° 124, OECD Publishing, Paris, doi.org/10.1787/d1b2b844-en.

Simkin, S. M. et al. (2016), "Conditional Vulnerability of Plant Diversity to Atmospheric Nitrogen Deposition Across the United States", *Proc Natl Acad Sci U S A*, 113(15), doi.org/10.1073/pnas.1515241113.

Simpson, D. (2015), "Present Day and Future Ozone in a Swedish, European and Global Perspective", presentation at a seminar on 'Ground-level Ozone – Present-day and Future Risks for Impacts on Vegetation', 9 March 2015, Stockholm.

Sinha, E. et al. (2017), "Eutrophication will Increase During the 21st Century as a Result of Precipitation Changes", *Science*, 357(6349), doi.org/10.1126/science.aan2409.

Sobota, D.J. et al. (2015), "Cost of Reactive Nitrogen Release from Human Activities to the Environment in the United States", *Environmental Research Letters*, 10(2), doi.org/10.1088/1748-9326/10/2/025006.

SRU (2015), *Nitrogen: Strategies for Resolving an Urgent Environmental Problem*, German Advisory Council on the Environment, Berlin.

Sutton, M.A. et al. (2011), *The European Nitrogen Assessment, Sources, Effects and Policy Perspectives*, Cambridge University Press, United Kingdom.

Swackhamer, D.L. et al. (2004), "Impacts of Atmospheric Pollutants on Aquatic Ecosystems", *Issues in Ecology*, N° 12, Ecological Society of America, Washington D.C.

USEPA (2017), "Review of the Primary National Ambient Air Quality Standards for Oxides of Nitrogen", Proposed Rules, *Federal Register*, 82(142), 26 July 2017, 82 FR 34792, gpo.gov/fdsys/pkg/FR-2017-07-26/pdf/2017-15591.pdf.

USEPA (2016a), *Integrated Science Assessment for Oxides of Nitrogen—Health Criteria (2016 Final Report)*, National Center for Environmental Assessment, Research Triangle Park, NC, EPA/600/R–15/068, epa.gov/ncea/isa/recordisplay.cfm?deid=310879.

USEPA (2016b), *Toxicological Review of Ammonia (Noncancer Inhalation)*, EPA/635/R-16/163Fa, Integrated Risk Information System National Center for Environmental Assessment, Office of Research and Development, U.S. Environmental Protection Agency Washington, DC.

USEPA (2015), "National Ambient Air Quality Standards for Ozone; Final Rule", *Federal Register*, 80(206), 26 October 2015, 80 FR 65291, gpo.gov/fdsys/pkg/FR-2015-10-26/pdf/2015-26594.pdf

USEPA (2013a), "National Ambient Air Quality Standards for Particulate Matter; Final Rule", Federal Register, 78(10), 15 January 2013, 78 FR 3086, gpo.gov/fdsys/pkg/FR-2013-01-15/pdf/2012-30946.pdf.

USEPA (2013b), "Final Aquatic Life Ambient Water Quality Criteria For Ammonia—Freshwater 2013", *Federal Register*, 78(163), 22 August 2013, 78 FR 52192, gpo.gov/fdsys/pkg/FR-2013-08-22/pdf/2013-20307.pdf.

USEPA-SAB (2011), *Reactive Nitrogen in the United States: An Analysis of Inputs, Flows, Consequences and Management Options*, EPA-SAB-11-013, U.S. Environmental Protection Agency's Science Advisory Board, Washington D.C.

Van Grinsven, H.J.M. et al. (2013), "Costs and Benefits of Nitrogen for Europe and Implications for Mitigation", *Environ. Sci. Technol.*, 47(8), doi.org/10.1021/es303804g.

Zhang, F. et al. (2015), "Cyanobacteria Blooms and Non-alcoholic Liver Disease: Evidence from a County Level Ecological Study in the United States", *Environmental Health*, 14(41), doi.org/10.1186/s12940-015-0026-7.

ORGANISATION FOR ECONOMIC CO-OPERATION AND DEVELOPMENT

The OECD is a unique forum where governments work together to address the economic, social and environmental challenges of globalisation. The OECD is also at the forefront of efforts to understand and to help governments respond to new developments and concerns, such as corporate governance, the information economy and the challenges of an ageing population. The Organisation provides a setting where governments can compare policy experiences, seek answers to common problems, identify good practice and work to co-ordinate domestic and international policies.

The OECD member countries are: Australia, Austria, Belgium, Canada, Chile, the Czech Republic, Denmark, Estonia, Finland, France, Germany, Greece, Hungary, Iceland, Ireland, Israel, Italy, Japan, Korea, Latvia, Lithuania, Luxembourg, Mexico, the Netherlands, New Zealand, Norway, Poland, Portugal, the Slovak Republic, Slovenia, Spain, Sweden, Switzerland, Turkey, the United Kingdom and the United States. The European Union takes part in the work of the OECD.

OECD Publishing disseminates widely the results of the Organisation's statistics gathering and research on economic, social and environmental issues, as well as the conventions, guidelines and standards agreed by its members.

OECD PUBLISHING, 2, rue André-Pascal, 75775 PARIS CEDEX 16
(97 2018 43 1 P) ISBN 978-92-64-30742-1 – 2018

Lightning Source UK Ltd.
Milton Keynes UK
UKHW030717011119
352727UK00006B/617/P